EVALUATING PROTEIN-CARBOHYDRATE INTERACTIONS INDUCED BY

MULTIVALENT CARBOHYDRATE-FUNCTIONALIZED DENDRIMERS

by

Kristian Henri Schlick

A dissertation submitted in partial fulfillment
of the requirements for the degree

of

Doctor of Philosophy

in

Biochemistry

MONTANA STATE UNIVERSITY
Bozeman, Montana

October 2010

UMI Number: 3433820

All rights reserved

INFORMATION TO ALL USERS
The quality of this reproduction is dependent upon the quality of the copy submitted.

In the unlikely event that the author did not send a complete manuscript
and there are missing pages, these will be noted. Also, if material had to be removed,
a note will indicate the deletion.

Dissertation Publishing

UMI 3433820
Copyright 2011 by ProQuest LLC.
All rights reserved. This edition of the work is protected against
unauthorized copying under Title 17, United States Code.

ProQuest LLC
789 East Eisenhower Parkway
P.O. Box 1346
Ann Arbor, MI 48106-1346

©COPYRIGHT

by

Kristian Henri Schlick

2010

All Rights Reserved

APPROVAL

of a dissertation submitted by

Kristian Henri Schlick

This dissertation has been read by each member of the dissertation committee and has been found to be satisfactory regarding content, English usage, format, citation, bibliographic style, and consistency and is ready for submission to the Division of Graduate Education.

Dr. Mary J. Cloninger

Approved for the Department of Chemistry

Dr. David Singel

Approved for the Division of Graduate Education

Dr. Carl A. Fox

STATEMENT OF PERMISSION TO USE

In presenting this dissertation in partial fulfillment of the requirements for a doctoral degree at Montana State University, I agree that the Library shall make it available to borrowers under rules of the Library. I further agree that copying of this dissertation is allowable only for scholarly purposes, consistent with "fair use" as prescribed in the U.S. Copyright Law. Requests for extensive copying or reproduction of this dissertation should be referred to ProQuest Information and Learning, 300 North Zeeb Road, Ann Arbor, Michigan 48106, to whom I have granted "the exclusive right to reproduce and distribute my dissertation in and from microform along with the non-exclusive right to reproduce and distribute my abstract in any format in whole or in part."

Kristian Henri Schlick

October 2010

ACKNOWLEDGEMENTS

First, I would like to thanks my advisor, Dr. Mary Cloninger for her guidance and support throughout the years. Her continued enthusiasm always provided a renewed vigor for chemistry, even when results seemed just out of reach. Thank you to Dr Greg Gillispie, with whom I had the opportunity for a great collaborative effort. I am grateful for the support of the members of my committee as well, being most helpful in the isolated cases where all other avenues of knowledge were exhausted. Of course I need to mention the awesome office staff, the controllers of our fate, especially Jennifer Smith, who could handle any task with an elegant poise.

My thanks to the members of the Cloninger group, both past and present, have always provided an excellent atmosphere to be around and it was the highest honor to be allowed to work with each of them. I have made some good friends in chemistry department, most notably Brandon and Enoch, who have helped me learn to live life outside the lab as well as within. Thanks to the friends that were there for me outside of the department as well; it was perhaps you that I needed most. You know who you are. And of course, thanks to my family for their ongoing support, even from across states and countries. Konrad and Kathryn, never give up, this world is yours.

TABLE OF CONTENTS

1. INTRODUCTION 1
   - Cell Surface Glycosylation 1
   - Protein-Carbohydrate Interactions 3
   - Lectins 5
   - Multivalent Binding 10
   - Dendrimers 14
   - Protein Aggregation 20
   - Summary 26

2. BINDING OF MANNOSE-FUNCTIONALIZED DENDRIMERS WITH PEA LECTIN AND CONCANAVALIN A 29
   - Introduction 29
   - Hemagglutination 32
   - Isothermal Titration Microcalorimetry 37
   - Precipitation Assays 39
   - Transmission Electron Microscopy 44
   - Conclusions 46
   - Experimentals 47

3. INHIBITION BINDING STUDIES OF GLYCODENDRIMER-LECTIN INTERACTIONS USING SURFACE PLASMON RESONANCE 50
   - Introduction 50
   - Self-Assembled Monolayers 53
   - Surface Plasmon Resonance 56
   - The Four Parameter Logistic Equation 67
   - Conclusions 68
   - Experimentals 70

4. TRISMANNOSE-FUNCTIONALIZED DENDRIMERS TO INVESTIGATE CLUSTERING EFFECTS BY ELISA 74
   - Introduction 74
   - Patterning and Clustering 74
   - Synthesis of Trismannose-Functionalized Dendrimers 76
   - Enzyme-Linked Immunosorbent Assays 81
   - Conclusions 88
   - Experimentals 89

TABLE OF CONTENTS - CONTINUED

5. CHARACTERIZATION OF PROTEIN AGGREGATION VIA INTRINSIC FLUORESCENCE LIFETIME 93

    Introduction 93
    Fluorescence Lifetime 95
    Con A Complexation Measured by Fluorescence Lifetime 97
    Conclusions 111
    Instrumentation 113
    Data Analysis of Fluorescence Lifetime Measurements 117
    Experimentals 127

6. CONCLUDING REMARKS 128

    APPENDIX A: Glycoproteomics with Boronic Acid Derivatives 132

REFERENCES CITED 172

LIST OF TABLES

Table		Page
2.1	Hemagglutination Inhibition Assays for Mannose-Functionalized Dendrimers	34
3.1	Comparison of Hemagglutination Inhibition Assays with SPR results for mannose-functionalized dendrimers with Concanavalin A	61
3.2	Curve fit parameters for inhibition experiments determined by GraphPad Prism 4	66
4.1	ELISA IC$_{50}$ values for tris-cluster dendrimers **17-20**	84
4.2	ELISA IC50 values for mannose functionalized dendrimers **3-6** and monovalent methyl-mannose	86
5.1	Calculated k_{obs} for compounds **2-4** and **6**	109
5.2	Kinetic data determined from global fits of k_{obs}	110
A.1	^{1}H-NMR titrations of **22** (15 mM) with fructose (150 mM)	144

LIST OF FIGURES

Figure		Page
1.1	Structures of A, B, and O blood group determinants	2
1.2	G_{M1} pentasaccharide bound to cholera toxin	5
1.3	Ca^{2+} and Mn^{2+} dependent Concanavalin A bound to α-D-methyl-mannose	8
1.4	Binding modes involved in multivalent interactions	11
1.5	Above: Assembly of an AB polymer (linear polymer). Below: Assembly of an AB_2 polymer (dendrimer)	15
1.6	Carbohydrate-sensing dendrimer	17
1.7	Aggresome formation model	23
2.1	Structures of (a) Con A and (b) pea lectin	30
2.2	Mannose-functionalized dendrimers **1-6**	31
2.3	Pea lectin (green) and Con A (orange) structures superimposed	36
2.4	ITC Profile of precipitating Con A and mannose functionalized G(5) PAMAM dendrimer	38
2.5	ITC profile of pea lectin (0.030 mM) with **5** (2 mM in sugar) at 27 °C	39
2.6	Precipitation assays with pea lectin	41
2.7	TEM images of a) 10 mM G5-man b) 10 mM G5-man treated with 57 mM pea lectin for ~20 hours (10x diluted)	45
3.1	Model of complex formation measured by surface plasmon resonance	51
3.2	A schematic representation of mannose functionalization of a gold surface	55
3.3	Top: Doubly-referenced sensorgram of Con A binding to the mannose-functionalized surface. Bottom: Affinity profile of Con A to the mannose-functionalized surface, fit using Scrubber 2	57

LIST OF FIGURES - CONTINUED

Figure		Page
3.4	(a) Mannose-functionalized poly(amidoamine) (PAMAM) dendrimers. (b) A schematic representation of the inhibition binding experiment	58
3.5	Inhibition of 2 μM Con A injected over the mannose-functionalized gold surface	59
3.6	Inhibition by mannose-functionalized PAMAM dendrimers of 2 μM Con A	60
3.7	Attempted fit of Con A to a kinetic 1:1 binding model	63
3.8	Inhibition by serial dilutions of galactose-functionalized G4 dendrimer	66
4.1	Trismannose-cluster-functionalized PAMAM dendrimer	76
4.2.	Spacer compound **7**, sugar cluster **8**	77
4.3	Mannose functionalized dendrimers **3-6**	78
4.4	SDS-PAGE of glycodendrimers visualized by a modified Periodic Acid-Schiff method on a 15% gel	80
4.5	ELISA Inhibition Procedure	82
4.6	ELISA inhibition graphs for compounds **17-20** and **3-6**	83
4.7	IC_{50} values for **17-20**	88
5.1	Mannose-functionalized PAMAM dendrimers	95
5.2	Jablonski diagram describing fluorescence	96
5.3	Fluorescence assay data for additions of G(2)-man **2** into 100ug/mL Con A	98
5.4	Fluorescence assay data for additions of G(3)-man **3** into 100ug/mL Con A	99
5.5	Fluorescence assay data for additions of G(4)-man **4** into 100ug/mL Con A	100
5.6	Fluorescence assay data for additions of G(6)-man **6** into 100ug/mL Con A	101

LIST OF FIGURES - CONTINUED

Figure		Page
5.7	Fluorescence assay data for additions of Me-man into 100ug/mL Con A	102
5.8	Fluorescence assay data for additions of G(4)-gal into 100ug/mL Con A	103
5.9	Complex formation upon addition of 17.2 mM Me-man, 16.3 µM G(4)-gal and 16.3 µM G(4)-man **4**	103
5.10	Glycodendrimer-mediated lectin aggregation	104
5.11	Structure of Con A with active binding sites highlighted in red and tryptophan residues highlighted in green	105
5.12	Complex formation of Con A with time as a result of dendrimer addition	106
5.13	Kinetic data for compounds **2-4** and **6**. k_{obs} presented on a mannose concentration basis	107
5.14	Kinetic data viewed in terms of dendrimer concentration	108
5.15	Diagram of Varian Eclipse Spectrometer, modified be Fluorescence Innovations, Inc. for fluorescence lifetime measurements	115
5.16	Schematic drawing of dye pumped laser configuration	116
5.17	Normalized free and complexed waveforms along with their difference scaled by a factor of 10	118
5.18	Overlap of complexed waveform for G2-G6 and standard deviation scaled by a factor of 100	119
5.19	Fit of 14.3 µM G(3)-man data for determination of k_{obs}	121
5.20	Fit of 1.8 µM G(3)-man data for determination of k_{obs}	121
5.21	Binding data of **2** fit to a 1:1 exponential association model	123
5.22	Binding data of **3** fit to a 1:1 exponential association model	124
5.23	Binding data of **4** fit to a 1:1 exponential association model	125

LIST OF FIGURES - CONTINUED

Figure   Page

5.24   Binding data of **6** fit to a 1:1 exponential association model..........................126

A.1   Glycosensor **21** based on boronic acid and benzophenone............................135

A.2   Absorbance profile of phenylboronic acid (PBA) binding to Alizarin Red S (ARS). Bottom: double reciprocal plot from which K_{eq} for the PBA-ARS interaction is determined..........................137

A.3   Absorbance profile of phenylboronic acid (PBA) binding to Alizarin Red S (ARS), inhibited by fructose additions. Bottom: plot from which K_{eq} for the PBA-fructose interaction is determined..........................139

A.4   Top: Absorbance profile of o-hydroxymethyl phenyl boronic acid binding to Alizarin Red S (ARS). Bottom: Inhibition by fructose additions...............140

A.5   Precursor glycosensor **22**..........................141

A.6 ^{1}H-NMR of o-hydroxymethyl phenylboronic acid before addition of fructose..........................143

A.7 ^{1}H-NMR of o-hydroxymethyl phenylboronic acid after addition of fructose..........................144

A.8   Double reciprocal plot to determine K_a of the boronic acid-fructose interaction..........................145

A.9 ^{1}H-NMR of **22** before addition of fructose..........................146

A.10 ^{1}H-NMR of **22** after addition of fructose..........................147

A.11   ITC profiles of a) 75 mM fructose, b) 75 mM sialic acid, and c) 75 mM Me-glcNAc injected into 5 mM boronic acid in 0.1 M PBS, pH 7.4. d) competitive ITC profile of 75 mM fructose injected into 5 mM boronic acid and 50 mM Me-glcNAc..........................149

A.12   ITC profiles after removing injections resulting in endothermic spikes and constraining N=1..........................151

A.13   Mass spectrometry of fetuin and **21**..........................153

LIST OF FIGURES - CONTINUED

Figure   Page

A.14 Mass spectrometry of *N*-Acetyl glucosamine and **21**, irradiated at 365 nm in water 154

A.15 Mass spectrometry of *N*-Acetyl glucosamine and **21**, control experiment in water 155

A.16 Alexa 555, benzophenone conjugated glycosensor **23** 156

A.17 Mass spectrometry of *N*-Acetyl glucosamine and **23**, control experiment in water 157

A.18 Mass spectrometry of *N*-Acetyl glucosamine and **23**, irradiated at 365 nm in water 158

A.19 Alexa 555 conjugated glycosensor **24** 159

A.20 Structures of microarray glycosides which consistently yielded strong fluorescence 161

LIST OF EQUATIONS

Equation		Page
3.1	The four parameter logistic equation	68
5.1	Linear combination of basis waveforms	118
5.2	Kinetic association binding for a one-phase association	120
5.3	Observed rate constant expressed as dependent on the rate of association and the rate of dissociation	122
A.1	Equation to determine K_a via the ARS Assay	164
A.2	Equation to determine K_a via competitive ITC	165
A.3	Ratio of free and complexed boronic acid peaks determined by ^{1}H-NMR Assay	167
A.4	Equation to determine K_a via ^{1}H-NMR Assay	167

LIST OF SCHEMES

Scheme		Page
4.1	Synthesis of deacetylated heterogeneously functionalized tris-mannose cluster and ethoxyethanol PAMAM dendrimers	78

ABSTRACT

Understanding protein-carbohydrate interactions is essential for elucidating biological pathways and cellular mechanisms but is often difficult due to the prevalence of multivalent interactions. A better understanding of the basic behavior of protein-carbohydrate interactions is critical for controlling cellular proliferation and recognition processes for novel therapeutic methods to be successful. Many procedures that exist for evaluating protein-carbohydrate interactions are often limited to monovalent interactions or small polymers. Given that many cellular processes, such as those attributed to the immune system, are enhanced multivalently or are aggregation-driven, there is a need to reveal the behavior and basic requirements for multivalent binding and aggregation.

Evaluating these interactions on large, multivalent scaffolds such as synthetically controllable dendrimers provides an important tool towards accurately determining the role of glycosylation in biological systems. Here, different approaches to measure the interactions of proteins with glycodendrimers are described, ranging from simple qualitative assays to novel quantitative methods of assessment. Quantitative methods such as Isothermal Titration Calorimetry and Surface Plasmon Resonance are severely limited when used with multivalent systems, and do not provide as accurate results as monovalent systems. When dealing with multivalent systems, inhibition assays often provide more reproducible results.

Through these experiments, it has become increasingly apparent that aggregates play a significant role in multivalent systems, and current methods to evaluate these interactions leave much room for improvement. Assay design is important both for basic identification and understanding of any interaction, especially higher-order interactions involving multivalency. Endgroup patterning and presentation was explored to determine their role in multivalent affinity enhancements. Using a novel fluorescence lifetime method, glycodendrimer-mediated aggregation was successfully characterized. The work here evaluates the effectiveness of assays used for carbohydrate interaction, translated to a multivalent scaffold, with special consideration to large-order aggregates.

Keywords: Dendrimer, Glycodendrimer, Carbohydrate, Protein-Carbohydrate Interactions, Multivalency, Aggregation, Concanavalin A, Assay Design, Fluorescence Lifetime

CHAPTER 1

INTRODUCTION

Cell Surface Glycosylation

Carbohydrates play an integral role in biological systems and come in a variety of forms such as glycoproteins, glycolipids, polysaccharides and monosaccharides. Glycosylation is the most common post-translational modification, with more than half of all proteins estimated to have one or more glycan chains.[1] Glycosylation serves a multitude of functions such as promoting correct protein folding and conferring stability upon a protein.[2] Additionally, many cells surfaces are heavily glycosylated which presents a first line of information about the cell to the outside world.

Cellular carbohydrates are mainly presented as glycoproteins or glycolipids that are able to recruit carbohydrate recognizing proteins.[3] As surface glycoconjugates, the carbohydrates are involved in a variety of processes such as fertilization, cell signaling, inflammatory responses, cancer metastasis, and the glycoconjugates serve as attachment sites for infectious bacteria and viruses, toxins, and hormones. For many pathogens, this recognition event is the first step in infection and communication with other cells.[3] The ABO blood group system uses sugars on the erythrocyte surface as the antigens to promote agglutination when paired with an incompatible blood type, which formed the basis for the development of safe transfusions (Figure 1.1).[4,5]

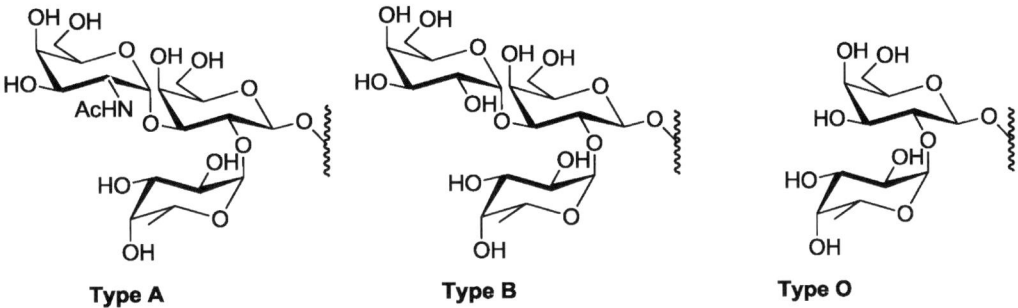

Figure 1.1. Structures of A, B, and O blood type determinants

The field of glycomics is rapidly growing and is rife with complexity and analytical challenges. Because of their numerous responses and the fact that many different structures can encode a single function, developing a structure-function relationship model for complex glycans is important to evaluating the glycome.[6] This will allow for an approach in many ways similar to proteomic and genomic efforts. The amount of possible glycans and glycan linkages make creation of a glycomic library a daunting task. Analysis of carbohydrate interactions is further compounded by heterogeneous glycosylation patterns present throughout biological systems, and is made even more complex by further modifications to the sugar such as methylation, phosphorylation, sulfation and acetylation, all of which are important to proper cellular signaling.

Currently the list of known glycans contains over 2000 structures and is constantly growing.[7] This variety provides the cell with an ability to encode massive amounts of information, but also poses significant analytical challenges to understanding the role of carbohydrates in biological systems. Glycomic analysis by mass spectrometry[8], lectin microarrays[9], and metabolic labeling of cellular glycans[10] are at the forefront of mapping the glycome and have been successfully used to identify sugar-based biomarkers of diseases[11]. However, a basic understanding of cell surface adhesion and binding is still needed, as are analytical methods to elucidate the requirements for cellular carbohydrate binding, particularly from a multivalent and aggregation perspective.

Protein-Carbohydrate Interactions

Protein-carbohydrate interactions on the cell surface play key roles in many cellular processes. The recruitment and activation of cells for mechanisms of inflammation and for the mounting of an immune response, for example, are protein-carbohydrate mediated events. Protein-carbohydrate interactions are also critical in other processes such as the infection of host cells by viruses and bacteria, the adhesion and metastatic spread of cancer cells, and even cellular differentiation and growth.[12, 13]

Reports of specific interactions between sugars and proteins can be traced back to Emil Fischer, who proposed a "lock and key" model to refer to enzymes which recognize specific carbohydrates.[14, 15] One of the first crystal structures to demonstrate the specific recognition of glycans by enzymes was lysozyme, a highly specific endoglycosidase capable of specifically cleaving β1-4 linkages in bacterial peptidoglycan.[16, 17] This eventually led to the 1945 Nobel prize received by Fleming, Chain, and Florey for their discovery of the antibacterial activity of lysozyme and penicillin. Since then, many other carbohydrate-binding proteins have been identified, such as Concanavalin A[18] and the Influenza Virus Hemagglutinin[19].

The energetics of protein-carbohydrate binding is the product of several contributing factors, such as the preferred conformation of a ligand or short-range interactions between the sugar and the residues in the binding site.[20] If the surface of a protein is complementary to the structure of the binding carbohydrate, water present in the protein binding pocket is displaced and rearranged around the bound complex. Solvation energies due to entropy are very large and cannot be reliably calculated for compounds that are as hydrophilic as sugars. Thus, even when the energetic contributions of van der Waals and hydrogen bonding interactions have been estimated, estimations of solvation energies can result in large errors, making overall energy calculations imprecise.

A consensus binding pattern of glycan-bound proteins has been solved by X-ray crystallography, showing that sugars are generally bound weakly in shallow pockets close to the protein surface, with typical K_d binding affinities in the millimolar to micromolar range. In some cases, however, the glycan is bound inside a cleft of the protein that is essentially inaccessible to the bulk solvent, and these carbohydrates can interact with hydrophobic residues, displaying K_d binding affinities ranging from 10^{-6} M – 10^{-10} M.[21] Cholera toxin is a pentameric adhesion protein which binds to the G_{M1} pentasaccharide in this fashion, with three residues making contact with the protein (Figure 1.2).[22, 23] The hydrophobic face of the terminal β-linked galactose residue also interacts with an aromatic side chain of the protein (Trp-88), and this interaction is typical of many glycan-protein complexes.

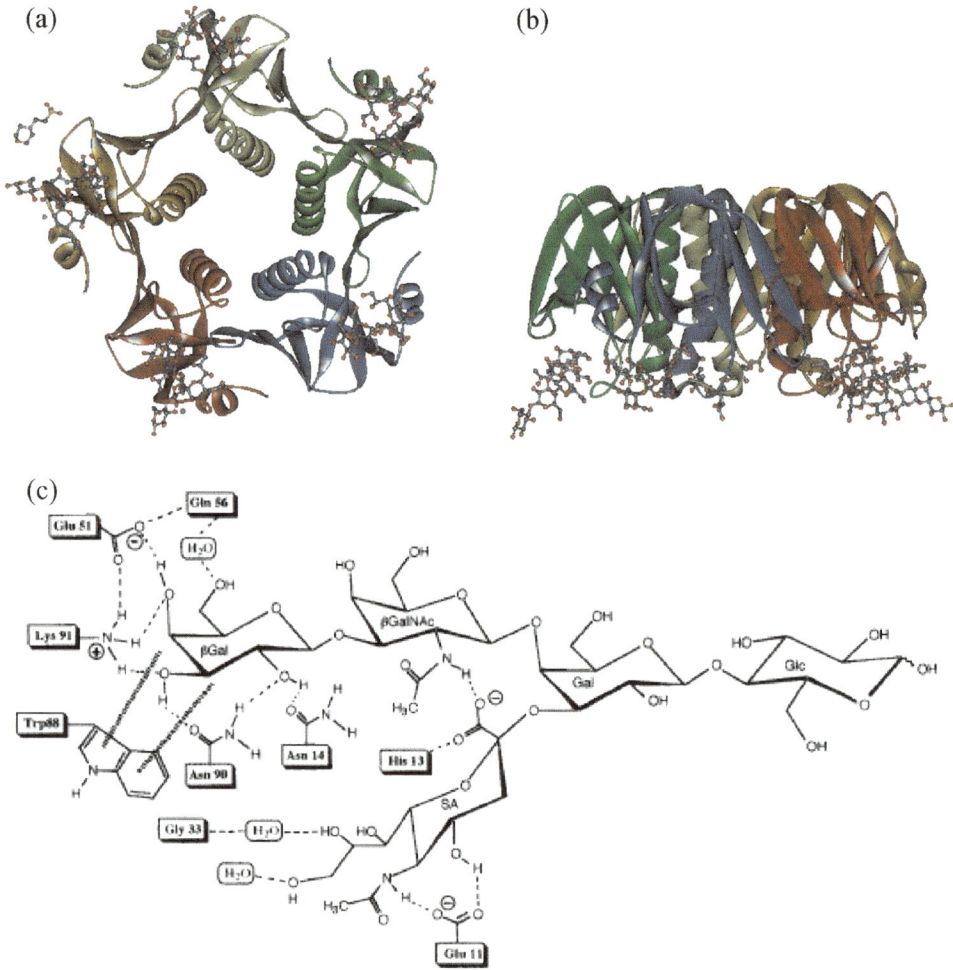

Figure 1.2. G_{M1} pentasaccharide bound to cholera toxin.[23] (a) View towards the binding surface.[23] (b) View perpendicular to the 5-fold axis, with the binding surface downward. (c) Simplified structure of G_{M1} binding to cholera toxin (cf. Reference 20).

Lectins

Lectins are proteins that recognize and bind specific carbohydrate epitopes. Lectins are defined as carbohydrate binding proteins that were not generated by the immune system and that lack enzymatic activity.[24] They have been described as interpreters of the "sugar code", as carbohydrates are able to far surpass amino acids and nucleotides in terms of information storing capacity and serve as ligands in biorecognition.[25] Like the fields of genomics and proteomics, the term *glycomics* has been coined as an analogous term to describe the systematic study of the glycan structures of an organism. Studying protein-carbohydrate interactions using lectins provides a valuable tool to understanding the surface patterning and modes of interaction of the glycans, which can in turn provide important information about information transfer, recognition, and infection.

Lectin activity was first documented in 1860 from the coagulation of red blood cells by rattlesnake venom: "one drop of venom was put on a slide and a drop of blood from a pigeon's wounded wing allowed to fall upon it. They were instantly mixed. Within three minutes the mass had coagulated firmly, and within ten it was of arterial redness."[26, 27] Hemagglutination activity was also instrumental to the discovery of the cell-bridging capacity of proteins in plant extracts, when the toxic extract of the castor bean (*Ricinus Communis*) caused the agglutination of red blood cells.[28] It was this discovery that plant extracts are a rich source of hemagglutinins that led to the purification of Concanavalin A (Con A) by crystallization and its subsequent role in glycomics.[29]

It was later found that several plant and animal hemagglutinins react with erythrocytes of different blood groups, displaying antibody-like selectivity, which led Boyd to introduce the term "lectin" in 1954: "It would appear to be a matter of semantics as to whether a substance not produced in response to an antigen should be called an antibody even though it is a protein and combines specifically with a certain antigen only. It might be better to have a different word for the substances and the present writer would like to propose the word *lectin* from Latin *lectus*, the past principle of *legere* meaning to pick, choose or select."[30] In its present definition, the term *lectin* has been expanded to include other proteins based on structural and functional similarities. Some toxins and monomeric carbohydrate binding proteins have been included to shift away from the experimental focus on agglutination, which requires at least bivalency to crosslink cells. Currently, there are three criteria for a protein to be considered as a member of the lectin family[26]:

1) Carbohydrate-binding activity
2) Distinction from immunoglobulins
3) Lack of enzymatic activity

Concanavalin A (Figure 1.3), as the first lectin to be successfully purified and crystallized from the jack bean, is the most commonly used plant lectin and is commercially available. As such, numerous research experiments have used Concanavalin A as a benchmark protein for carbohydrate binding studies. The lectin is a homotetramer at biological pH with a molecular weight of 26500 g/mol, with each monomer unit able to specifically bind mannose or glucose residues.[31] Concanavalin A requires both Ca^{2+} and Mn^{2+} metal ions for carbohydrate binding, classifying it as a C-type lectin.

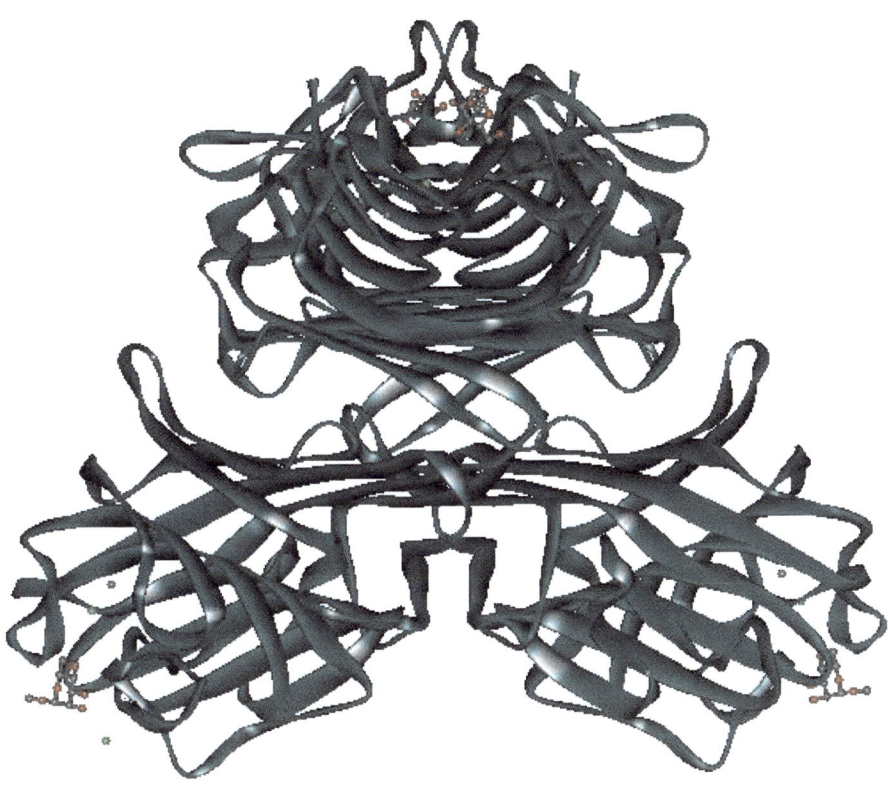

Figure 1.3. Ca^{2+} and Mn^{2+} dependent Concanavalin A bound to α-D-methyl-mannose.[32]

Lectins were first classified based on which carbohydrates they bind, then later on sequence homology and evolutionary relatedness.[33, 20] One group of lectins required calcium to bind and was therefore called C-type lectins, while another group required free thiols for stability and was called S-type lectins (these were later renamed galectins, as not all of them were thiol dependent but recognized β-galactosidases). P-type lectins were found by sequencing homologous two lectins that recognized mannose-6-phosphate. Although P-type and S-type lectins seem to recognize only a single class of sugars, others like C-type lectins are able to recognize a variety of sugars. Through information obtained from crystal structures, it has turned out that several groups of plant lectins have structural or sequence similarities to animal lectins.

Some biologically relevant lectins have been found in recent years, including Cyanovirin N, a lectin with a molecular weight of 11 kDa, originally isolated from cyanobacteria.[34] Cyanovirin N is able to prevent infection from the Human Immunodeficiency Virus (HIV) by binding gp120, a heavily glycosylated HIV envelope protein. Galectin-3 is another important lectin, found in humans. It is a 31 kDa monomer with a 14 kDa carbohydrate recognition domain which is able to oligomerize at higher concentrations through use of its tail-like domain.[35] Galectin-3 has been found to play significant roles in a number of disease processes, such as cell growth, cancer, immunity, and inflammation, including the promotion of fibrosis.[36, 37] Scarring is a natural consequence of injury or inflammation; galectin-3 activates fibroblasts responsible for the formation of connective tissue involved in the healing process. However, higher levels of galectin-3 have been reported in tumor cells, and have been linked to higher metastatic rates in cancer.[38, 39]

Among the many classes of lectins, the defining feature of many lectins is that they bind carbohydrates reversibly and non-covalently with weak affinities typically in the millimolar to micromolar range.[40] Since many biologically relevant interactions occur in the nanomolar range, it might seem at first that lectin-carbohydrate interactions are not biologically significant; however, lectins often increase their carbohydrate binding affinity through multivalent interactions.

Multivalent Binding

Sugars generally bind in shallow grooves close to the protein surface with dissociation constants in the millimolar to micromolar range. Because protein-carbohydrate interactions are of great importance for many biological events, multivalent interactions are often used by nature to overcome weak interactions and to increase the binding avidity.[41, 42] Many lectins provide an excellent system for the study of multivalent interactions, as they generally have multiple binding sites relatively distant from each other (3-7 nm), and bind carbohydrates with a high specificity.[43]

Presenting a carbohydrate ligand in a multivalent fashion to a protein often enhances binding beyond what could be expected from the increase in ligand concentration. This phenomenon is called the "cluster effect" or "multivalent effect".[40, 44] There are two requirements in order to get such an affinity enhancement: first, the lectin has to have more than one binding site, and second, the ligand must present multiple glycoside ligands in the correct orientation and spacing.

Multivalent interactions can occur in a variety of binding motifs (Figure 1.4). Simple multivalent enhancement effects can be described by a statistical or proximity effect arising from when multiple ligands are closely clustered around a receptor. In response to this, receptor clustering may occur. Bivalent and higher interactions occur when multiple binding sites are occupied by a multi-dentate ligand. Under the right circumstances, higher order aggregates and crosslinking can occur in cases where multivalent ligands are able to span receptors located on multiple carbohydrate-binding proteins, depending on the concentration and binding strength of the interaction.

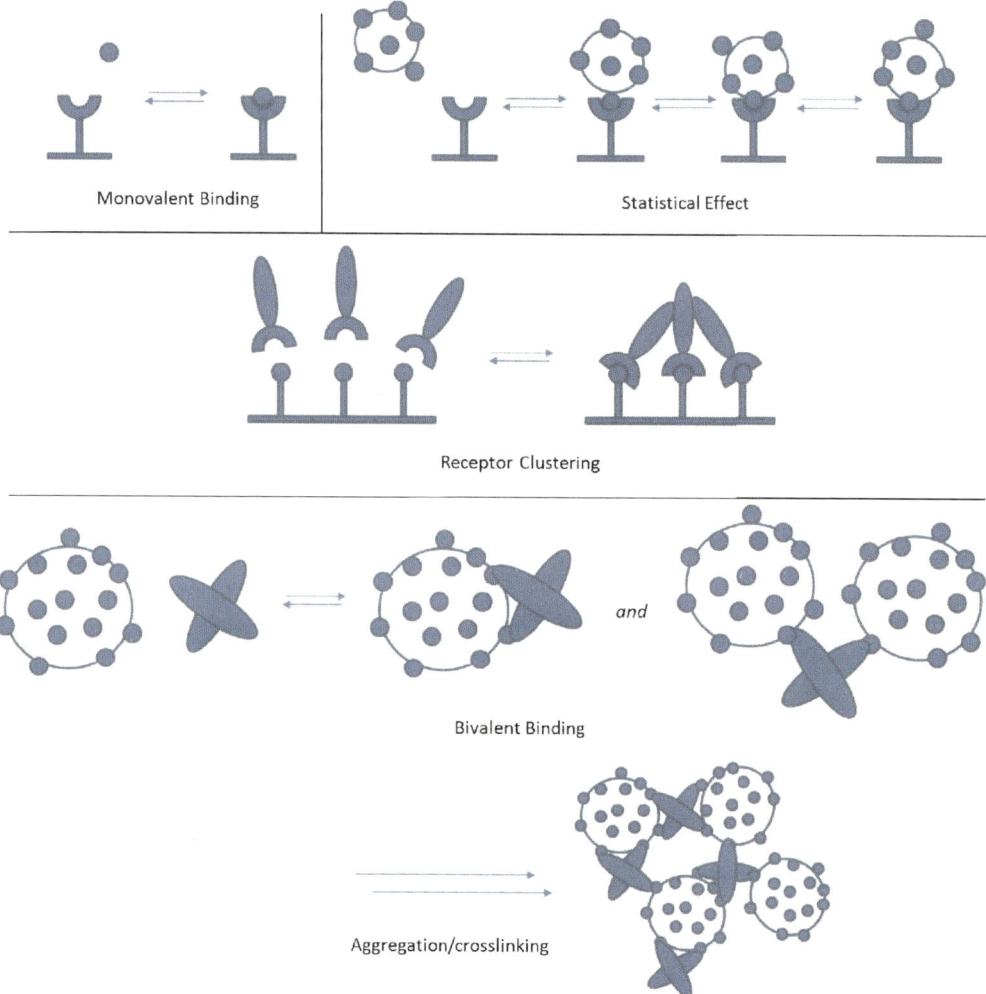

Figure 1.4. Binding modes involved in multivalent interactions (adapted from reference 45).

Designing synthetic multivalent ligands requires less emphasis on determining the optimal binding ligand for a protein than would be required for optimization of a monovalent interaction. This greatly reduces the amount of synthetic exploration that is required, such as the synthesis of deoxy analogs. Instead, emphasis is placed on designing the proper presentation and flexibility of the ligands[46]. Selection of both the scaffold and the spacer is important, and affects solubility of the multivalent ligand. For biologically relevant interactions, the compound must be water soluble as well as nontoxic to the cell. Cytotoxicity of the compound can be markedly decreased by selection of an appropriate spacer compound; for example, polyamidoamine (PAMAM) dendrimers have been shown to display a marked decrease in cytotoxicity after functionalization with a polyethyleneglycol (PEG) spacer.[47]

Spacer length influences the ability of a compound to undergo multivalent binding and can have influences on the binding behavior of the receptor-ligand interaction. Several studies have found that compounds with their ligands optimally distributed allow for better inhibition, greatly influencing the strength of binding.[48, 49] A coumpound whose ligands are located too closely cannot undergo multivalent binding, and ligands that are that are spaced too far apart induce a larger entropic cost upon binding.

As with spacer length, the valency of the system also has an optimum. Depending on the presentation of the ligands and the nature of the carbohydrate-binding protein used, maximum binding potency can be achieved as a trivalent cluster[50], or may require as many as 50 repeating units to be an effective inhibitor[51]. As the size of the multivalent display changes, its optimal binding properties and optimal ligand valency change and can be further customized using multiple ligand types.[52, 53] This allows for a large degree of tunability when using multivalent ligands that is far beyond what can be achieved using monovalent ligands.

Because multivalency plays such an important role in biological recognition events, understanding multivalent biological systems on the molecular level can suggest strategies for the design and application of novel drugs. Synthetic multivalent molecules can be specifically designed to inhibit or promote biological processes, greatly improving the current scope of drug design.

Artificial scaffolds are often functionalized with carbohydrates to study protein-carbohydrate interactions and are characterized using lectins.[54] However, quantitation of these interactions has been difficult, and additional assays are needed. Coupling protein-carbohydrate interactions to a readily quantifiable measurement, such as fluorescence, would allow for researchers to fine-tune synthetic compounds for maximum efficiency. Furthermore, characterizing protein-carbohydrate interactions at the membrane level would provide biologically more relevant data than interactions occurring in free solution. Once the parameters necessary for protein-carbohydrate binding are elucidated and quantified, interactions such as bacterial and viral adhesion can be efficiently controlled and understood, as well as other cell processes such as cell growth and differentiation. A thorough understanding of parameters governing protein-carbohydrate interactions is necessary if therapeutic agents relying on protein-carbohydrate interactions are to be developed.

Dendrimers

Dendrimers are a class of branched macromolecules with very well-defined chemical structures. They are created in a cascade synthesis of reactions which allows for precise control of the dendrimers' size, shape and properties. By controlling the number and nature of the tethered functional groups, the solubility and reactivity of the molecule can be customized to suit biological conditions.[55] Unlike other polymers, dendrimers have a low polydispersity, though higher generations are not perfectly controlled; with more iterations, a branch may be omitted leading to fewer than the theoretical active chemical groups on the surface. Nevertheless, dendrimers are ideal probes of components involved in protein-carbohydrate interactions due to their well-defined characteristics. Using dendrimers large enough to span the lectins' binding sites enables us to probe multivalent effects, as well as other biologically relevant events.

The word *dendrimer* comes from Greek word *dendron*, meaning tree or branch, referring to the branched nature of the molecule. Although dendrimer chemistry first emerged in 1978[56], it wasn't until the 1990's when the number of publications in the research field dramatically increased.[57, 58, 59] Dendrimers distinguish themselves from traditional linear polymers in two critical ways.[60] First, they are formed from AB_n monomers, resulting in hyperbranched molecules, as opposed to AB monomers which produce linear polymers (Figure 1.5). Secondly, dendrimers are synthesized in an iterative fashion which incorporates a more controlled number of monomer units with each successive step. For example, an AB_2 polymer would roughly double the number of monomeric units incorporated, whereas an AB_3 polymer would roughly triple the number. Each iterative step in the process leads to the addition of one more layer of branches, or endgroups, to the dendrimer scaffold, called a generation. Therefore, the generation of a dendrimer is a measure of the number of repetition cycles performed, and an indication of its size and number of endgroups present.

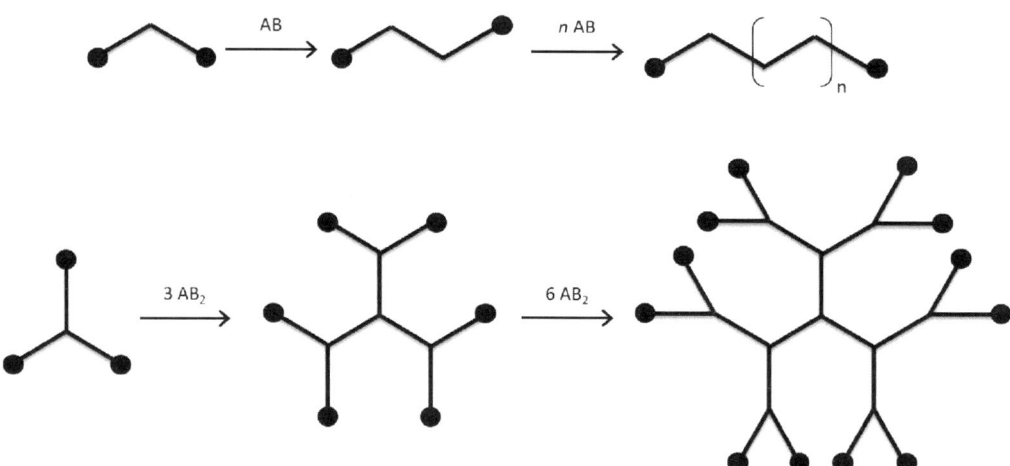

Figure 1.5. Above: Assembly of an AB polymer (linear polymer). Below: Assembly of an AB_2 polymer (dendrimer). (Adapted from reference 61)

Dendrimers can be synthesized using two main strategies: a divergent approach and a convergent approach. In the divergent approach, pioneered by Newkome[62] and Tomalia[63], dendrimers are synthesized from the dendrimer core outward in a stepwise fashion. Each cycle has a number of reactive groups n which can react with n monomer units, resulting in an additional layer/generation. In the next cycle, $2n$ reactive groups are present for an AB_2 monomer, or $3n$ for an AB_3 monomer, dependent on the monomeric unit's branch multiplicity (Figure 1.5). In a convergent approach, the periphery of the dendrimer is constructed first, and then coupled to the dendrimer core. This reduces the amount of reactions happening at any given time, as the outer dendritic structures are coupled to less reactive sites on the focal point. In general, this results in a more homogenous product than dendrimers prepared by a divergent approach, as defects begin to accumulate with the large amount of coupling reactions inherent in divergent syntheses. A classic example of this convergent approach is work reported by Fréchet[64], where the synthesis of dendritic "wedges" is described, then coupled to a central core. This allows for more control per generation growth step while minimizing the possibility of failed coupling sequences attributed to each divergent step.

The high amount of control inherent in dendrimer synthesis allows them to be used in a wide array of applications. Dendrimers are able to adopt a more defined three dimensional structure as opposed to linear polymers, which often assume a random-coil conformation. Higher generation dendrimers often adopt a spherical three-dimensional form which closely resembles the structure of a globular protein. As such, dendrimers can be used as highly controlled, synthetic protein mimics offering a good first approximation to proteins which are potentially difficult to isolate. Using a dendrimer as a scaffold, elucidating molecular recognition events becomes a much more controllable process, and specific interactions can be more easily targeted.

A common way to use the unique architecture of a dendrimer is to functionalize the outer endgroups with a molecule of interest. Dendrimers can contain a large number of functional groups on their surface, making them an attractive target for interactions where a close proximity of a large number of ligands is important. Several groups have used this approach to study a number of applications. Shinkai *et al.* have used dendrimers containing boronic acids and anthracene groups as a sensor to report on the binding events of carbohydrates (Figure 1.6).[65] Upon binding of the carbohydrate to the boronic acid moiety, the anthracene experiences a change in fluorescence, which can be monitored.

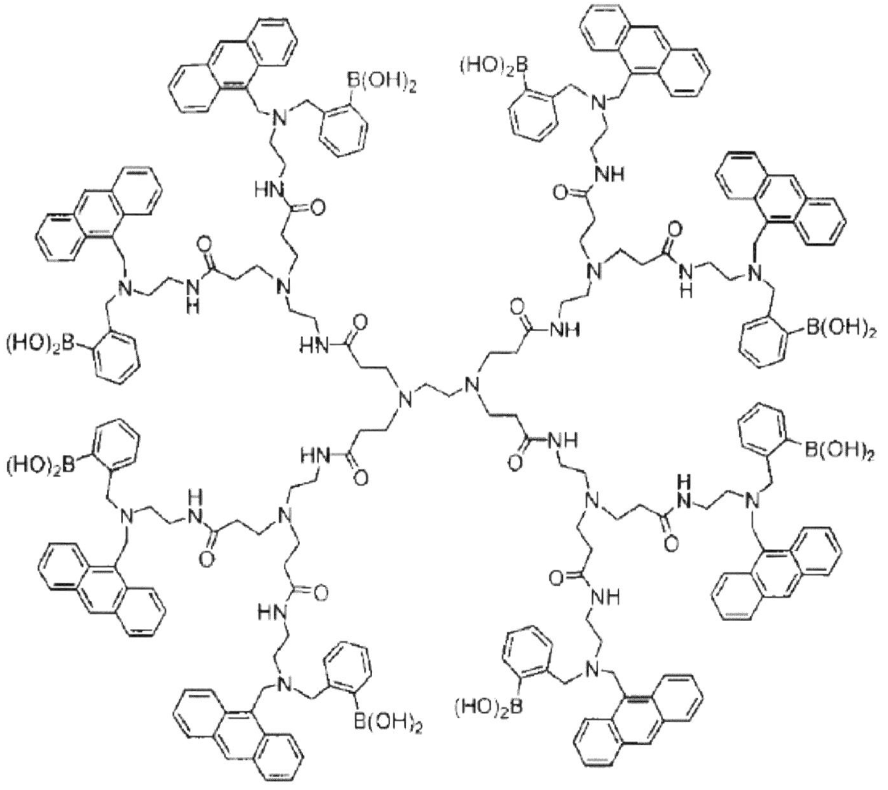

Figure 1.6. Carbohydrate-sensing dendrimer. The boronic acids are able to bind to carbohydrates, upon which the anthracene fluorescence changes (cf. Ref 65).

In our group, this strategy of surface functionalization has been extensively used to investigate the binding behavior of carbohydrates with various proteins (cf. Chapters

2-5). The multivalent nature of the dendrimer along with its high tunability allows for elucidation of basic interactions which are not yet well understood. A synthetic scaffold provides a means to determine the binding behavior of an interaction using a more controlled environment, which may not be possible when probing complex biological interactions.

Surface functionalization of dendrimers has great potential in the medical field as well, with potential applications in areas such as targeted drug delivery. The surface of the dendrimer can be decorated to bind to specific recognition domains, directing the molecule to specific sites in the body. Because of the large amount of endgroups on the dendrimer, a drug can be attached to the dendrimer, delivering the drug to only the specific sites targeted. A further advantage is that otherwise insoluble drugs can be delivered in this way, as the surface and the interior of the dendrimer can be modified to solubilize otherwise hydrophobic molecules.

In addition to having unique surface properties, the interior of the dendrimer has been used in a wide range of applications as well. The interior of the dendrimer is often its own microenvironment, protected from the bulk solvent by the dendrimer surface. This imparts the dendrimer with additional useful qualities. Instead of coupling a drug to the exterior surface of a dendrimer, the interior of dendrimers often have ample space to accommodate guest molecules. When a dendrimer is comprised of a hydrophobic interior and charged surface functional groups, its overall structure resembles that of a micelle.[61] As opposed to normal micelles, these dendrimers possess concentration-independent micelle properties and do not disassemble below a characteristic critical micelle concentration. Several examples of such host-guest interactions exist, exhibiting the monomeric nature of the created dendrimers as well as their ability to effectively encapsulate a guest molecule at a wide range of concentrations.[66, 67]

The core of a dendritic structure can have as much of an effect as the interior branches and is just as versatile. Moore *et al.* synthesized conjugated dendrimer systems containing a perylene core and investigated their fluorescence properties.[68] The branches were found to absorb the energy from light, then transferred the energy to the perylene core. The energy-cascade ability of these dendrimers increased with the number of peripheral sites which were augmented by increasing the dendrimer generation. In a similar experiment, this approach was used to cause photoisomerization of a central azobenzene unit by absorbing low-energy photons channeled to the core.[69]

Although unfunctionalized polyamidoamine (PAMAM) dendrimers have been shown to be cytotoxic, these properties can be controlled through choice of linker compounds and functionalization of the dendrimer endgroups. In addition to controlling the physical properties of the PAMAM dendrimers, its biochemical properties such as immunogenicity can also be controlled by surface functionalization.[70] Because of the dendrimer's highly tunable physical and biochemical properties and ideal size range, PAMAM dendrimers serve as excellent tools for elucidating biologically relevant multivalent interactions.

Protein Aggregation

One prominent issue in studying multivalent interactions is the problem with extensive cross-linking and aggregation. Cross-linking occurs when a multivalent ligand can bind to two different proteins with multiple binding sites. If these proteins then bind to another multivalent ligand, followed by another, large complexes can form which may become insoluble and form a precipitate. This makes accurate determination of kinetic and thermodynamic binding events difficult, as precipitation and complex formation contribute to the overall energetics of the system, incorrectly reflecting aggregation phenomena.

Given that a large number of biological events are mediated by multivalent interactions, multimeric assembly is important for biological function as well. Many systems induce multimerization of their receptors, leading to biologically relevant processes such as signal transduction.[71] Some cross-linking interactions lead to unique macromolecular assemblies, such as signaling apoptosis induced by the clustering and separation of receptors.[72]

Hydrophobicity plays a large role in protein aggregation and clustering phenomena. In general, proteins contain these assembly-promoting residues not spread out over an entire sequence but rather contained in well-defined regions of a sequence.[73] In this respect, protein assembly is different from protein folding, where residues responsible for forming the folded protein structure are relatively distant from each other in the sequence.[74]

During folding, proteins that are only partially folded can leave unwanted hydrophobic domains exposed, leading to inappropriate associations and protein aggregation. Protein aggregation is toxic to cells and interferes with the ability of polypeptides to fold correctly, especially in cases where the high cytosolic concentration of macromolecules causes a significant "crowding effect".[75] The effects of protein aggregation are amplified by the fact that the entire protein needs to first be fully synthesized before stable folding can occur.

Ultimately, a protein either becomes correctly folded or aggregates. This is dependent on a number of factors such as pH, temperature, ionic strength, redox environment, mutations, or simply translational errors.[75] Because misfolded proteins cannot be entirely avoided, cells have adapted various mechanisms to minimize misfolding and eliminate misfolded proteins before they aggregate.[76]

The unfolded protein response pathway regulates proteins with the aid of the endoplasmic reticulum (ER) through ER chaperones which bind and stabilize exposed hydrophobic residues.[77] Chaperones are less important for small, monomeric proteins, but are critically important and often required for larger proteins to fold correctly. An ER stress response pathway activates once the amount of misfolded proteins exceeds the folding capacity of the ER, inducing ER chaperones to limit the protein synthesis to a level manageable by the ER.

Proteins unable to fold properly are targeted for degradation by the ubiquitin-proteasome system, a multisubunit complex in the cytosol and nucleus which mediates the degradation of proteins into smaller peptides.[75] Misfolded secretory and transmembrane proteins are retained in the ER and are returned to the cytosol, where they are cleaved by the proteasome. When the misfolded or unfolded proteins accumulate to the point that the proteasome degradation pathway cannot keep up, the incorrectly formed proteins form aggregates which are then targeted by the aggresome pathway.

Proteins are generated by a polyribosomal system, reading the mRNA coding for a given number of peptides. Proteins not able to fold correctly from the polysome co-aggregate to form aggresomal particles of a uniform size. Once formed, these aggresomal particles are transported by the microtubule organizing center (MTOC) to a large cellular "garbage bin-like" structure known as the aggresome.[78] The aggresome could also be described as an aggregate of aggregates whose movements are an active process governed by microtubules and a dynein motor. Chaperones, ubiquitination enzymes, and proteases are recruited to the aggresome, which facilitate the clearance of the aggregated proteins. Ultimately, intermediate filaments reorganize to form a 'cage' surrounding the aggresome and is engulfed by autophagosomes, which then fuse to lysosomes, resulting in the degradation of the remaining protein matter by lysosomal hydrolases (Figure 1.6). In a sense, the aggresome can be viewed as an "obese degrasome", an exaggeration of a normal cellular process in which aggregated proteins are cleared by the degrasome.[75]

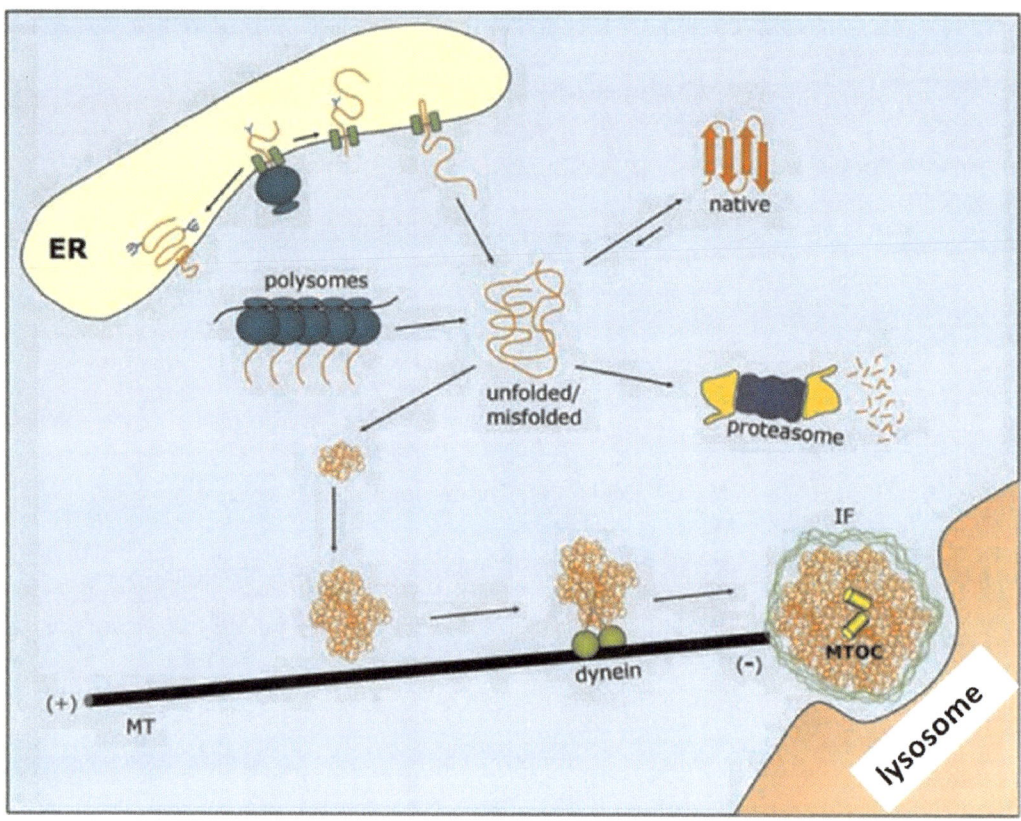

Figure 1.7. Aggresome formation model. Unfolded or misfolded proteins can originate from translating polysomes, proteins failing the endoplasmic reticulum's "quality control" mechanism, or proteins damaged by stress. If the unfolded/misfolded proteins fail to fold correctly and are not degraded by the proteasome, they can form aggregates throughout the cells. These aggregates are transported by microtubules to the MTOC in a process dependent on a dynein motor complex. The intermediate filaments cytoskeleton (IF) reorganize to form a 'cage' surrounding the aggresome, then is transported to the lysosome for degradation.[75]

Cells possess cellular signaling mechanisms to promote the formation of aggresomes, as the sequestration of aggregated proteins improves their clearance.[79] The intermediate filament (IF) cytoskeleton rearranges to form a cage around the aggresome.[80] Although the molecular requirements of IF rearrangement are not well understood, dramatic changes in IF architecture have been linked to neurodegenerative diseases. Aggresomes have been linked to numerous diseases, and mutant forms of aggregation-prone proteins such as superoxide dismutase[81], α-synuclein[82], or huntington[83] leads to the formation of large aggregates in neurons and neurodegeneration. These neurodegeneration pathways mimic the pathology of Familial Amyotrophic Lateral Sclerosis (FALS), Parkinson's disease and Huntington's disease and time-course studies strongly correlate aggresome formation to behavioral abnormalities and cellular degeneration.[84, 85] It is the formation of aggresomes, not of aggregated protein, that suggests cytotoxicity, and coincides with motor neuron dysfunction and cell death.[86]

Although cell death in response to aggresome formation is not well understood, several mechanisms have been proposed. The appearance of aggresomes may cause cellular components to co-aggregate, resulting in reduced cell viability.[87] Cellular chaperones recruited to the aggresome to help in aggregated protein clearance may also adversely affect cell viability, since sequestering these chaperones would lead to a decrease in efficient folding for the rest of the cell.[75] Since transport of the aggresome is microtubule-dependent, cell death could be caused by inhibition of microtubule-based transport. Excessive association of the aggresome with the microtubules can override the neuron's ability for movement of essential components to cell extremities.[75, 88] In general, accumulation of aggregated proteins and persistent aggresomes leads to an induced cellular stress response, activating an apoptotic pathway. However, more research is required to explore the pathogenic mechanisms involved in the correlation between the appearance of aggresomes and the pathology associated with aggresomal diseases.

Protein aggregation can come as a direct result of the increased production of cellular proteins, and an overexpression of certain genes. Genetic variation between two individuals is often the result of a difference in regulatory protein variants or copy number of proteins, and is less the effect of coding sequence.[89] However, only a handful of genes are pathological when overexpressed. Most genes have no severe effect on growth when overexpressed, yet only a select subset of genes is dosage-sensitive and can cause a major phenotypic change.[90] With the overexpression of proteins, there is an increased amount of misfolded proteins present in the cell, which can lead to increased protein promiscuity. Disordered regions of misfolded proteins are prone to making promiscuous molecular interactions when their concentration is increased, leading to potentially lethal phenotypes.[91]

As such, aggregated proteins provide an excellent marker for a multitude of diseases. Controlling and identifying basic aggregation mechanisms however, is not yet well understood and requires further investigation. A number of amyloid diseases and neurodegenerative diseases arise from aggregated proteins, yet these diseases are at the present difficult to diagnose, especially in early stages. Several genetic markers have been identified to provide a measure statistical risk to individuals, but these markers do not provide information as to when or even if those at risk may expect the onset of disease.[92] Neuroimaging can be used to assess at-risk individuals, and this technique is displaying a significant advancement in the recognition of relevant proteins. Generally, the neuroimaging marker must be hydrophobic enough to exhibit high blood-brain barrier permeability and exhibit selective binding to amyloid aggregates yet not so hydrophobic as to cause aggregates themselves.[93] Clinical diagnosis of these diseases at an early stage is typically very inaccurate and difficult, as early clinical signs can often overlap with normal signs (e.g. aging).[92] In fact, significant neuroimaging and pathological changes do not always occur until later stages of the disease. Biochemical markers taken from a spinal tap or blood serum have been proposed but have not always been consistent or have produced a low sensitivity and specificity.[92, 94, 95]

Research into correctly diagnosing diseases, particularly diseases associated with protein aggregation mechanisms, continues to be a major focus of clinical therapeutics. Protein aggregation and the misfolding of proteins can be linked to the origin of many conformational diseases which can be either genetic or spontaneous. The proteins involved can either have an unstructured or linear unfolded form such as in Alzheimer's and Parkinson's disease, or Type II Diabetes, or can be globular, showing a folded 3D-structure. The basic requirements leading to the formation of aggregates are still not well understood, yet aggregation has a large impact on the structure and function of cells and cell signaling. For this reason, studying multivalently-induced aggregation behavior can yield a deeper understanding of how diseases may proliferate, and can provide a mechanistically-based approach to underlying processes responsible to diseases.

Summary

Understanding multivalent protein-carbohydrate interactions has implications for elucidating the biological role of cell surface glycosylation. As a common post-translational modification, glycosylation holds important roles in complex biological processes. Many of these processes improve binding via multivalent interactions often leading to additional binding modes such as cross-linking and aggregation. The following report provides insight into various methods of detecting these binding modes using carbohydrate-functionalized dendrimers.

The first project focuses on binding differences between two similar lectins. Concanavalin A is a tetramer and pea lectin is a dimer, but both lectins recognize the same carbohydrate ligands, and pea lectin has a weaker affinity for its ligands. Several techniques are described to elucidate the binding interaction differences, such as isothermal titration calorimetry (ITC), hemagglutination assays, and precipitation assays.

The next project evaluates the use of surface plasmon resonance (SPR) as a high-throughput system to measure multivalent protein-carbohydrate interactions through inhibition assays. SPR allows for automated binding studies using a minimal amount of materials over a carbohydrate-functionalized surface. This enables a comparison of solution-based binding studies with presentation of ligands on a surface.

The third project focuses on examining the pattern effects of carbohydrate ligands for evaluating proximity and multivalent effects. The added synthetic control allows for comparisons of cluster-functionalized dendrimers and randomly functionalized dendrimers, evaluating the effect tighter clustering has on multivalent binding.

The fourth project uses fluorescence lifetime to follow Concanavalin A binding multivalently in solution. This experiment makes use of a novel fluorescence spectrophotometer to track miniscule binding changes in real time and with high fidelity. This technique allows for the extrapolation of kinetic data without the need for extensive labeling procedures. Furthermore, this technique allows for the measurement of aggregation events, and enables the study of binding despite extensive aggregation that is occurring in the solution.

CHAPTER 2

BINDING OF MANNOSE-FUNCTIONALIZED DENDRIMERS WITH PEA (PISUM SATIVUM) LECTIN

Introduction

Among the lectins, legume lectins represent the best-characterized group of lectins. Legume lectins are typically homodimers or homotetramers that contain one highly conserved sugar-binding site on each monomeric unit.[96, 97, 98] Concanavalin A (Con A)[31, 99] and a lectin isolated from *Pisum Sativum* (pea lectin)[100, 101] are two such lectins. In solution, Con A is a homotetramer at biological pH, while pea lectin is a homodimer. Both proteins bind methyl mannose with specificity, although Con A has fourfold higher affinity than pea lectin for methyl mannose.[102] As shown in Figure 2.1, both Con A and pea lectin have binding sites about 65 Å apart.[32, 103]

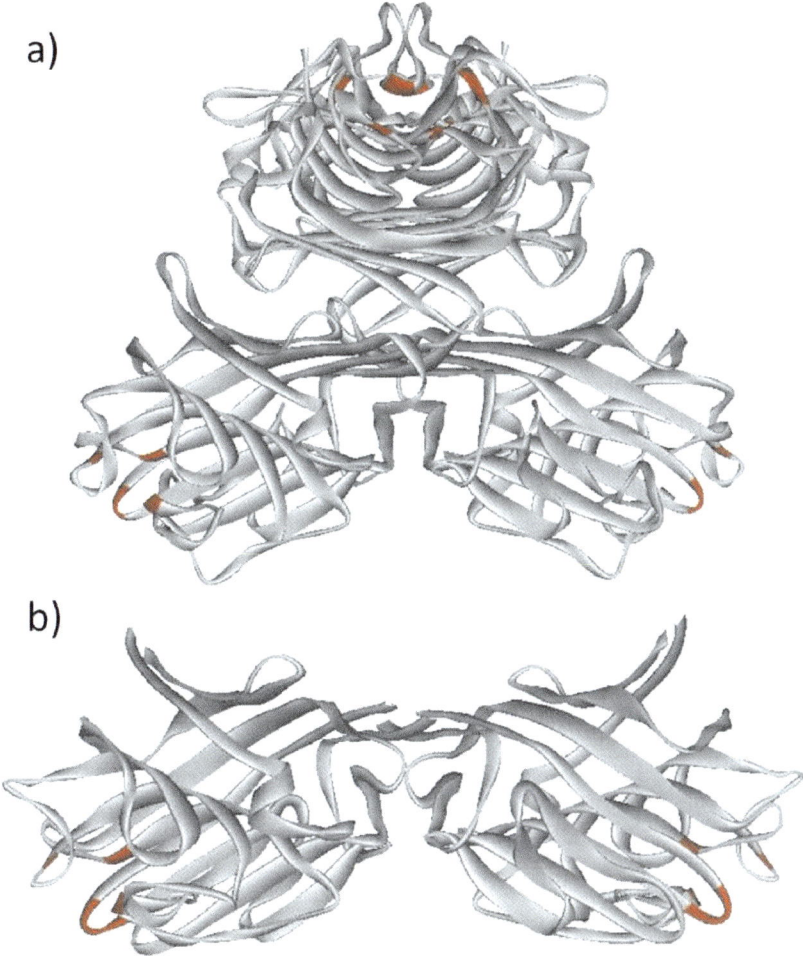

Figure 2.1. Structures of (a) Con A[32] and (b) pea lectin[101]. The binding sites of both lectins[99, 103] are about 65 Å apart; binding residues of both lectins are indicated in red.

Because lectin-carbohydrate adhesion generally involves multivalent interactions, a variety of glycopolymers that can span multiple lectin binding sites have been developed to help study these processes.[54] Dendrimers, for example, can be functionalized with sugars to study protein-carbohydrate interactions.[104] Understanding multivalent biological systems on the molecular level can suggest strategies for novel drugs. Synthetic multivalent molecules can be designed to either inhibit or promote biological processes, greatly improving drug design.

Unlike most polymers, dendrimers have a regular branching pattern with predictable physical properties.[55] By controlling the number and nature of the tethered functional groups, the solubility and reactivity of the molecule can be customized to suit biological conditions. The synthesis and lectin-binding properties of generation one through generation six mannose-functionalized PAMAM dendrimers (compounds **1–6**, Figure 2.2) with Con A have been previously reported.[105] In these Con A-dendrimer interactions, multivalent binding motifs were observed for large dendrimers (generations four through six, compounds **4–6**), while smaller statistical or proximity-type enhancements in binding were observed for **3**, and relative affinities comparable to methyl mannose were observed for dendrimers **1** and **2**.[106] Changing the size of the dendrimer framework (by changing the generation of PAMAM used) changed the type of interaction that was observed with Con A.

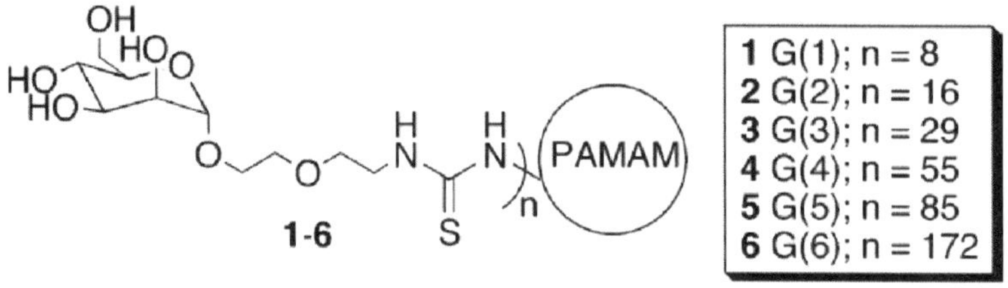

Figure 2.2. Mannose-functionalized dendrimers **1-6**.

In order to test the generality of the results that were obtained with Con A, binding studies of mannose-functionalized dendrimers to pea lectin were performed. The binding results for mannose-functionalized dendrimers with pea lectin are presented here and are compared to previously obtained results with Con A. Hemagglutination assays, precipitation assays, and isothermal titration microcalorimetry (ITC) studies are described.

Multivalent protein-carbohydrate interactions serve a critical function in many intercellular recognition events. A thorough understanding of their fundamental requirements is essential if therapeutic agents are to be developed that rely on protein-carbohydrate interactions. Through careful exploration of glycodendrimer interactions with legume lectins, our goal is to develop generally applicable parameters for development of therapeutic agents.

Hemagglutination Assays

Table 2.1 shows the relative activity of dendrimers **1–6** for pea lectin (concentration adjusted) compared to methyl-mannose. For comparison, Con A-dendrimer affinities are also shown.[52, 105] Although significant increases in affinity (on a per mannose basis) were observed with Con A as the generation of the dendrimer was increased, the affinity of the dendrimers for pea lectin remains roughly constant for the six generations of dendrimers.

Mannose-functionalized G(1) and G(2)-PAMAMs **1** and **2** were bound to Con A with comparable concentration adjusted affinities to that of methyl mannose, suggesting that monovalent binding between Con A and these dendrimers occurs. Higher generation mannose-functionalized PAMAMs, however, suggest glycoside clustering/proximity effects (G(3)-PAMAM **3**) and multivalent binding (G(4)- to G(6)-PAMAM **4–6**).[105, 106] No increase in binding is observed with pea lectin, which suggests exclusive monovalent binding. On a per mannose basis, mannose-functionalized G(2)- to G(6)-PAMAMs **2–6** actually bind with lower affinity than G(1)-PAMAM **1** or methyl mannose does.

Hemagglutination assays with Con A were performed with 18 μg/mL of Con A,[52] but assays with pea lectin were performed with only 2 μg/mL. The lower concentration of pea lectin was required since, for a low affinity lectin, free lectin is always detectable in solution at higher concentrations. Using a low concentration of pea lectin eliminated the background agglutination. Inhibiting concentrations of dendrimers and methyl mannose are shown in Table 2.1. To allow for comparison of the results for dendrimers **1–6**, methyl mannose was assigned the relative activity value of one with each protein, even though the inhibiting concentrations of Con A with methyl mannose and of pea lectin with methyl mannose are different.

Table 2.1. Hemagglutination Inhibition Assays for Mannose-Functionalized Dendrimers. Each reported value represents at least three assays.

	no. of sugars	**Con A** inhibiting sugar conc.[a] (mg/mL)	**Con A** rel. act./ mannose[a]	**Pea Lectin** inhibiting sugar conc.[b] (mg/mL)	**Pea Lectin** rel. act./ mannose[b]
Methyl mannose	1	2.5	1	1.25	1
1	8	6.2	1	2.5	1.3
2	16	4.4 ± 0.29	1.5 ± 0.1	5	0.67
3	29	0.34 ± 0.081	21 ± 5	10	0.35
4	55	0.035 ± 0.015	204 ± 90	10	0.36
5	95	0.024 ± 0.016	305 ± 209	10	0.37
6	178	0.021 ± 0.0053	354 ± 89	10	0.38

[a] Values are taken from references 52 (**1–2**) and 105 (**3–6**).
[b] Values were consistent under the conditions used, so no ranges are reported.

Presumably, unfavorable steric interactions between the dendrimer and pea lectin preclude bivalent binding. Con A has a concave surface that nicely compliments the roughly spherical shape of dendrimers **4–6**, but the area between the binding sites on pea lectin is flatter and therefore perhaps is less ideal for accommodating the dendrimer's bulk. Since pea lectin lacks the concave surface feature present in Con A, pea lectin may be unable to accommodate a multivalent binding motif for dendrimers **4–6**. Although they are clearly large enough to span the distance between mannose binding sites on pea lectin, **4–6** may be unable to bind in a bidentate fashion due to lack of shape complementarity.

A picture with the crystal structures of pea lectin and Con A superimposed upon one another reveals that the two proteins are quite similar (Figure 2.3). The only significant difference appears to be that Con A has larger loop regions that jut out from the protein, giving Con A a more curved shape than pea lectin between the binding sites. Although the difference seems relatively minor, such effects have previously been attributed to significant changes in binding motifs.[25] The affinity of pea lectin for methyl mannose is only about four times lower than the affinity of Con A for methyl mannose,[14] and it has been previously shown that small differences in affinity do not change the binding motif from bivalent to monovalent binding for Con A/dendrimer systems.[27] Also, the pea lectin dimer and the Con A tetramers are both stable entities at neutral pH, so motifs where the pea lectin dissociates to form monomers (which are unable to bind bivalently) seem highly unlikely.[107] Thus, it seems most likely that the monovalent binding pattern indicated by the hemagglutination assay results with pea lectin and dendrimers **4-6** is caused by the lack of shape complementarity between the two systems.

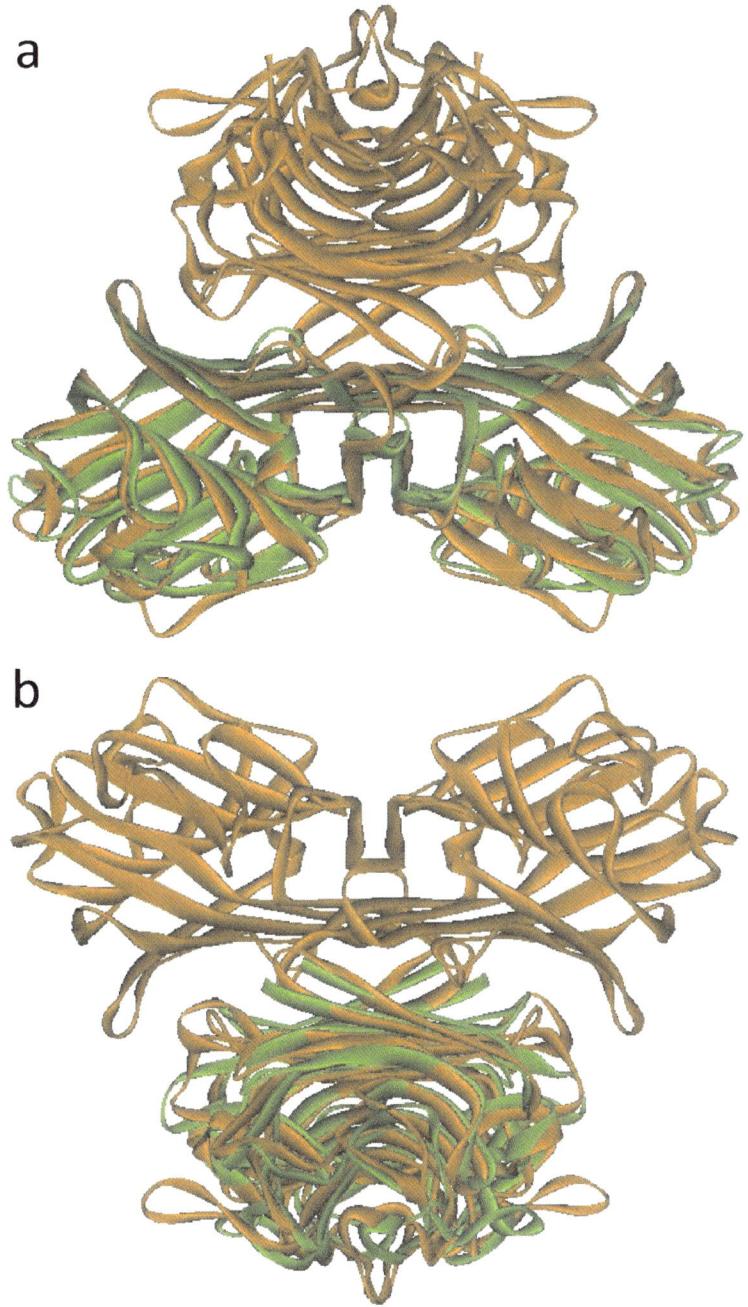

Figure 2.3. Pea lectin (green) and Con A (orange) structures superimposed. (a) Proteins are shown in the same orientation as in Figure 1 and (b) proteins are rotated 90°.

Isothermal Titration Calorimetry

Although hemagglutination assays provide helpful information about the relative activity of the mannose-functionalized dendrimers for Con A and pea lectin, they do not provide association constants or energetics of binding. Isothermal Titration Calorimetry (ITC) can provide valuable information about the number of binding sites (n), the enthalpy of binding (ΔH), and the association constant (K_a), from which the free energy of binding (ΔG) and the entropy of binding (ΔS) can be calculated. The binding of monosaccharides[33, 108] and small mannose-functionalized multivalent frameworks[109] to Con A and pea lectin have been previously reported, which led us to hypothesize that ITC might be used for binding studies with our dendrimers **1–6** as well.

Unfortunately, aggregation was repeatedly observed upon addition of the mannose-functionalized dendrimers **1–6** to Con A (Figure 2.4). Although pea lectin has a significantly lower affinity for methyl-mannose and causes less aggregation than Con A, precipitation was still observed. Upon inspection after addition of dendrimer precipitation could be visually seen in the adiabatic cell. Figure 2.5 shows a representative experiment during which minimal aggregation was observed. While an overall sigmoidal curve was obtained, a significant amount of noise is present. This is partially due to the relatively low amounts of protein used. In ITC measurements, the quantity $c = K_a M_t(0)$, where $M_t(0)$ is the initial macromolecule concentration, is important for optimum curve fitting to be achieved.[110] However, accurate ITC measurements require the presence of soluble complexes for the duration of the experiment. In order to minimize aggregation, a lower than optimal macromolecule concentration was required, which resulted in a low signal. The larger amount of variability at the beginning of the titration is mostly due to the formation of an aggregate. Because of the formation of an aggregate, the thermodynamic data obtained from the titration cannot be analyzed in detail.

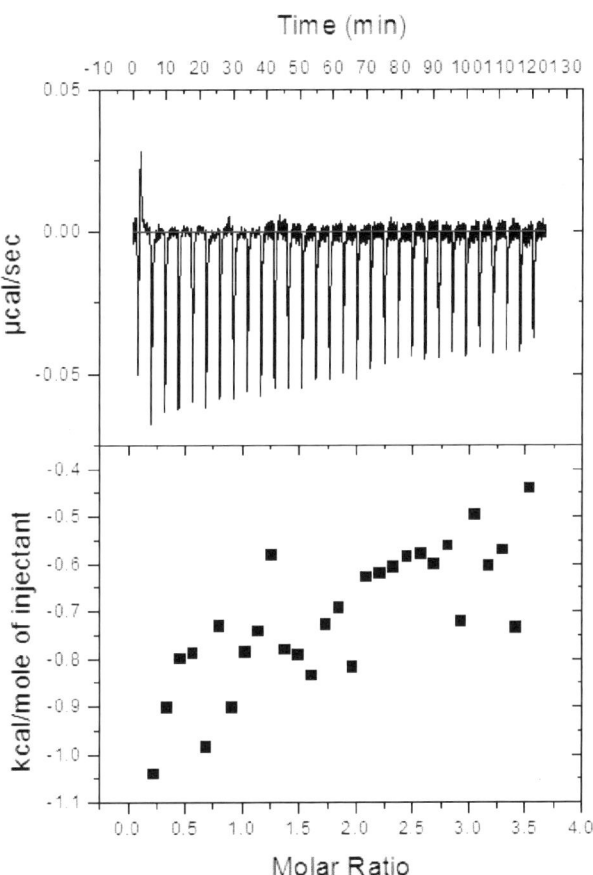

Figure 2.4. ITC Profile of precipitating Con A and mannose functionalized G(5) PAMAM dendrimer.

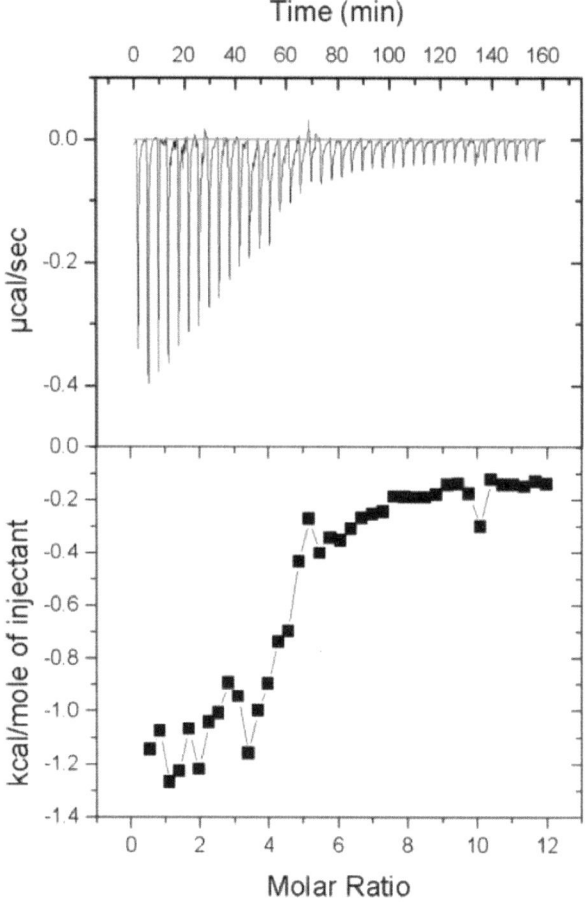

Figure 2.5. ITC profile of pea lectin (0.030 mM) with **5** (2 mM in sugar) at 27 °C. *Top*, data obtained for 40 automatic injections; *bottom*, integrated curve showing experimental points and the best fit.

Precipitation Assays

To further investigate aggregation behavior, precipitation assays were performed with pea lectin and dendrimers **1–6**. By treating varying concentrations of dendrimers with a constant amount of protein, the dendrimer-lectin binding stoichiometry is obtained.[111] If the concentration of pea lectin is high enough, the dendrimer will precipitate the protein. The ratio of dendrimer to protein at the point of maximum precipitation is considered the maximum number of pea lectins that are recruited by each dendrimer.

The ratio of mannose-functionalized dendrimer to pea lectin in the precipitates quickly increased with increasing amount of dendrimer until a maximum was reached, then the ratio steadily decreased (Figure 2.6). This behavior was different from our previously obtained results with Con A, where the ratio of protein to dendrimer remained fairly constant after a maximum had been reached.[52] When the assay was performed in water, as opposed to buffer, results resembled those for Con A. This change in behavior was likely caused by a change in the ionic strength of the solution.

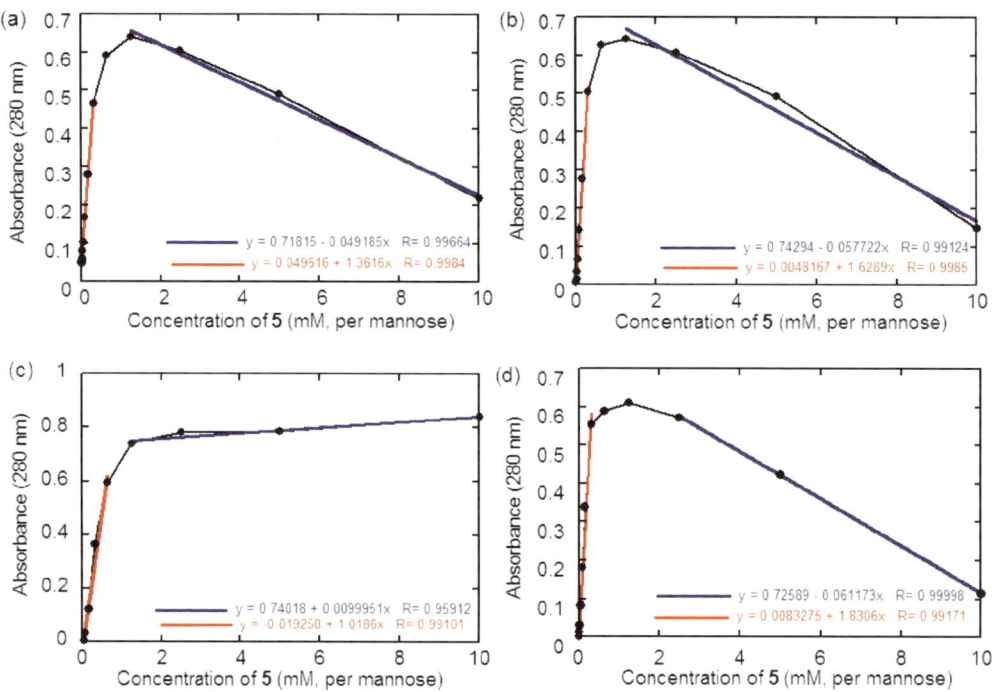

Figure 2.6. Precipitation Assays in a) 10 mM PBS, pH 7.2, adding 67 µM pea lectin, b) 0.1 M TBS, pH 7.4, adding 71 µM pea lectin, c) water, adding 71 µM pea lectin and d) 0.15 M NaCl, adding 60 µM pea lectin.

Similar effects were observed when Okada and co-workers studied the behavior of dendrimer-ion complexes.[112] Although Okada's study did not directly involve protein-carbohydrate interactions, parallels can be drawn for multivalent receptor-ligand interactions. Presumably, the dendrimer-protein complex became more soluble at high ionic strength due to the screening effect of microsalts weakening the binding interaction.[113] Such a screening effect could occur if salts interfere with electrostatic interactions and hydrogen bonding in the binding pocket, which would increase the off-rate for complex dissociation and would disrupt the larger cross-linked complexes. The formation of soluble protein-dendrimer complexes in the presence of high concentrations of dendrimer, as well as at low ligand concentrations, is possible if the ionic strength of the solution is sufficiently high, because only small aggregates would form under these conditions. Near the equivalence point, insoluble complexes are formed in extensive lattices.[41]

The precipitation assay results reported here are consistent with the aggregation behavior that was observed in ITC experiments, where the most noise occurs as the experiment approaches the maximum dendrimer to protein stoichiometry. That the salt content influences sugar binding by pea lectin was also shown by comparing hemagglutination assays with methyl mannose in buffers containing 0.15 and 0.35 M NaCl. Increasing the salt content from 0.15 to 0.35 M caused a 1.3-fold increase in the amount of methyl mannose required to inhibit agglutination. The results with methyl mannose suggest that a high salt content in the assay buffer can lower the affinity of the protein for mannose, which might disrupt the protein/dendrimer crosslinks and reduce the amount of precipitate observed in the precipitation assays.

The precipitation profiles in Figure 2.6 show that, for a mannose-functionalized G(5)-PAMAM dendrimer **5**, precipitation steadily increases, reaches a maximum, and then steadily decreases in high ionic strength buffers (Figures 2.6a,b, and d). In water, the precipitation profile levels off (Figure 2.6c). The point of intersection of the initial slope and the final slope are similar for all of the buffers tested, corresponding to an approximately 1:7 ratio of dendrimer **5** to pea lectin monomer.

Physical models suggest that the dendrimer could accommodate 11 pea lectin dimers, assuming bivalent binding. While the dendrimer : pea lectin ratio could theoretically arise from an average of bivalently binding three or four pea lectins, presumably extensive cross-linking is occurring, which is greatly weakened by the salting effects, thus forming a soluble complex at both high and low concentrations of dendrimer.[112, 113]

Overall, the precipitation assays seem to suggest that even a large dendrimer such as **5**, which is theoretically able to span binding sites 65 Å apart and experimentally was shown to bind multivalently to Con A, probably binds pea lectin in a monodentate fashion in high ionic strength buffers. This is consistent with the results obtained from the hemagglutination assays. The lack of shape complementarity between pea lectin and **5** may make multivalent binding less likely than with Con A; large cross-linking complexes probably occur more readily, which high ionic strength buffers can minimize. The shape complementarity between Con A and dendrimer is not present with pea lectin, and steric hindrance when binding dendrimer **5** to pea lectin may prevent multivalent binding motifs from occurring. Brewer and co-workers have previously presented a similar steric argument, in that the binding valency of a glycoprotein (which can be compared to our mannose-functionalized dendrimers) is influenced by the quarternary structures of the lectin and the interacting glycoprotein.[111]

Transmission Electron Microscopy

Because the stoichiometry obtained from the precipitation assays in water can suggest either bivalent or monovalent binding, TEM images were taken to elucidate the binding behavior in water (Figure 2.7). Figure 2.7a shows the glycodendrimer before addition of protein, which is a largely homogeneous solution. 24 hours after addition of protein, figure 2.7b shows large, irregular particles. This suggests that pea lectin binds to mannose-functionalized G5-PAMAM monovalently, forming extensive cross-linked complexes. In water, it is proposed that these complexes aggregate with each other due to non-specific binding interactions, which leads to continued precipitation under high glycodendrimer concentrations in low ionic strength buffers. In high ionic strength buffers, only the large cross-linked complexes near the equivalence point precipitate, as non-specific interactions are considerably weakened due to the screening effect of microsalts.[113]

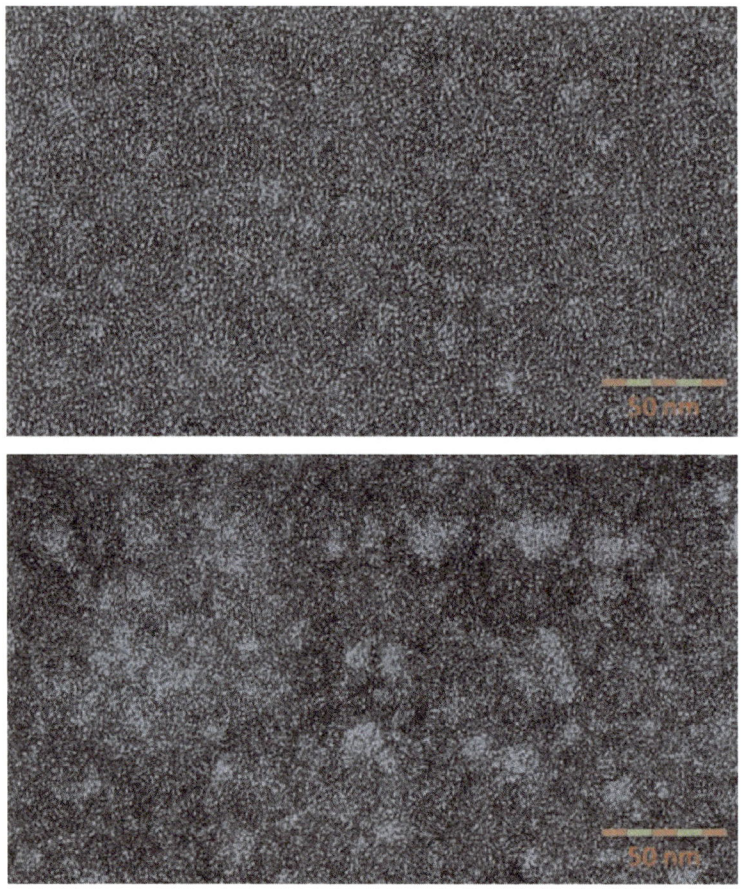

Figure 2.7. TEM images of a) 10 mM G5-man b) 10 mM G5-man treated with 57 mM pea lectin for ~20 hours (10x diluted).

The results of the TEM images confirm the previous results of hemagglutination and precipitation that pea lectin is bound monovalently by G5-PAMAM glycodendrimers. Although the glycodendrimer is theoretically able to span both binding sites, as is seen in our previous results with Con A, pea lectin binds mannose with a lower affinity necessary to enable bivalent binding.

Conclusions

The hemagglutination assays and precipitation assays described here suggest that mannose-functionalized PAMAM dendrimers **1–6** bind to pea lectin in a monovalent fashion, which contrasts with our previous results obtained using Con A. Hemagglutination assays for **1–6** with pea lectin showed no increase in binding affinity (on a per mannose basis) with increasing dendrimer generation. Previous experiments with Con A revealed significant increases in binding affinity as the dendrimer size increased, and the affinity changes were attributed to multivalent binding and glycoside clustering. Unfortunately, kinetic data for pea lectin/dendrimer binding could not be obtained via ITC due to aggregation. Precipitation profiles suggested monovalent binding of pea lectin by all dendrimers in high ionic strength buffers and in water.

Size and shape complementarity of the carbohydrate-functionalized dendrimers with the lectin appears to influence the multivalent (Con A/dendrimer) versus monovalent (pea lectin/dendrimer) binding motif. While the possibility exists that highly dynamic frameworks such as PAMAM dendrimers can undergo conformational shifts to overcome unfavorable steric repulsions and to achieve bivalent binding interactions, the unfavorable shape complementarity in the dendrimer/pea lectin system appears to induce monovalent association. The association is relatively weak (a micromolar dissociation constant is reported for pea lectin with methyl mannose),[102] and so higher affinity bivalent binding evidently cannot occur by overcoming the unoptimized shape complementarity in this case.

The comparative binding behaviors of the legume lectins, Con A and pea lectin, to carbohydrate-functionalized dendrimers **1–6** indicate that glycodendrimers may be very appropriate macromolecular therapeutic agents for targeted multivalent binding with some lectins but may be unsuccessful in other cases. The shape complementarity of the lectin with the carbohydrate-coated framework must be considered when dendrimer based systems are applied.

Experimentals

General Methods. Mannose-functionalized dendrimers **1–6** were synthesized as previously described in reference 52. Pea lectin was isolated from commercial green split peas according to a procedure from Trowbridge.[100]

Hemagglutination Inhibition Assay. Hemagglutination assays were performed similarly to previously published procedures.[114] The concentration of pea lectin was kept at 2 μg/mL for all assays. Assay buffer consisted of 0.5% w/v BSA in 10 mM phosphate buffered saline (PBS), pH 7.2. The pea lectin was added to serial dilutions of a 20 mg/mL glycodendrimer stock solution, and the solutions were incubated for 3 hours at room temperature. Rabbit erythrocytes (2% v/v in 0.5% w/v BSA) were added, and the lowest amount of dendrimer to cause inhibition was determined. Salt effects for monovalent binding of methyl-mannose to pea lectin were determined for 0.15 M, 0.25 M, and 0.35 M NaCl solutions.

Isothermal Titration Calorimetry. ITC experiments were performed similarly to previously published procedures[108, 110] using a microcalorimeter from Microcal, Inc. (Northhampton, MA). Injections of 6 µL of mannose-functionalized PAMAM dendrimer were added via a 250 µL syringe at an interval of 4 minutes into 1.435 mL of a pea lectin solution while stirring at 310 rpm. The concentration of the lectin ranged from 20 µM to 200 µM, and the sugar concentration ranged from 0.5 mM to 20 mM. Titrations were done at 27 °C in a 10 mM phosphate buffered saline solution (pH 7.2).

Precipitation Assay. Precipitation assays were performed using a modification of the procedure reported by Brewer and co-workers.[111] Serial dilutions of a 20 mM solution of mannose-functionalized G(5)-PAMAM were added to microcentrifuge tubes. A final volume of 500 µL was used. A pea lectin solution (500 µL, 60 to 71 µM as shown in Figure 2.6) was added to each tube and allowed to precipitate for ~20 h at room temperature. The supernatant was removed after centrifugation at 5000 rpm for 5 min. The pellet was then washed three times with 500 µL of cold buffer and dissolved in 2 mM methyl α-D-mannopyranoside to a final volume of 1 mL. The solutions were then analyzed for protein content using $A^{1\%}_{280} = 15.0$ for pea lectin.[115]

Molecular Modeling. Commercially available air-drying clay was used to model both the dendrimers and the pea lectin using a scale of 1 Å = 1/64 inch. **5** was modeled as a sphere with a 95/64 inches (= 95 Å) diameter. Pea lectin was modeled as a cylinder with a diameter of 29/32 inch (= 58 Å) and a length of 41/32 inches (= 82 Å).

Transmission Electron Microscopy. Images were taken on a LEO 912AB transmission electron microscope at an acceleration voltage of 100 kV and a 1 second exposure time. Samples from the precipitation assays were prepared by dropping a 5 μL spot onto a carbon grid, followed by a 5 μL spot of a staining solution of uranyl acetate. Some samples were diluted prior to staining to better visualize images.

CHAPTER 3

INHIBITION BINDING STUDIES OF GLYCODENDRIMER-LECTIN
INTERACTIONS USING SURFACE PLASMON RESONANCE

Introduction

Traditionally, hemagglutination and precipitation techniques have been used to determine relative interactions occurring between proteins and carbohydrates.[111, 114] However, these assays are limited in that they do not provide either thermodynamic or kinetic data about the protein-carbohydrate interactions. Isothermal titration microcalorimetry (ITC)[116] directly measures thermodynamic binding parameters but requires large amounts of both the ligand and the receptor to obtain good signal. The requisite conditions of ITC have the potential to promote aggregation conditions, especially when using large, highly multivalent scaffolds such as dendrimers.

Surface plasmon resonance (SPR) technology provides a valuable technique for accurately analyzing protein-carbohydrate interactions. This method has gained a great deal of interest in recent years, and has been used to analyze many ligand-ligate complexes.[117, 118] While many methods require the use of tags such as radiolabels, SPR eliminates the need for labeled reagents. In addition, when qualitative binding comparisons across a series of compounds are needed, the need for extensive purification protocols is reduced when specific ligand/receptor binding is involved. The ligand of interest is passed through the system at a constant flowrate and non-participating compounds are quickly washed away. SPR allows one to monitor complex formations occurring in real time, offering insights into the kinetics and mechanism of a reaction by measuring the change of the surface's refractive index upon binding (Figure 3.1).[117] The assay is sensitive enough that low affinity interactions can be detected, since measurements can be made in the presence of excess, unbound protein. This is especially useful in protein-carbohydrate studies, where the interactions are known to be weak.

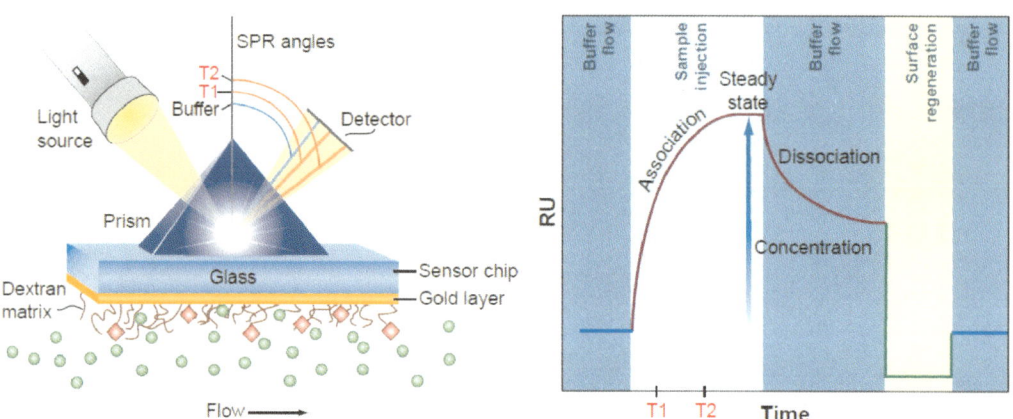

Figure 3.1. Model of complex formation measured by surface plasmon resonance. The SPR optical unit and a sensor chip detects molecules in the flow solution (green spheres), which passes by the target molecule linked to the dextran matrix (pink diamonds). Time points T_1 and T_2, shown in the schematic sensorgram (right), correspond to the two red SPR angles, which shifts as molecules are bound to the surface. The complex dissociates upon reintroduction of the buffer. The response to a regeneration solution will fall below baseline if its refractive index is lower than that of the buffer.[119]

The use of SPR to elucidate binding events has seen a significant increase in recent years. Several noteworthy examples of the application of SPR to the study of protein-carbohydrate interactions have been reported; summaries of a few studies with direct relevance to this work follow. Keusgen and co-workers, for example, recently reported a lectin screen using immobilized oligosaccharides in SPR. The goal of this work was to devise a system that would ultimately allow for the identification of new lectins from natural sources.[120] Gabius, Kamerling, and co-workers immobilized glycoproteins for sensitive detection of galectin-1 and *Ricinus communis* agglutinin in solutions containing mixtures of proteins[121], and Nishimura and co-workers used SPR to evaluate how effectively sialyl lexis X functionalized b-cyclodextrins inhibited the binding of E-selectin to a SLeXn-BSA functionalized sensor chip[122]. Suda and co-workers recently compared the binding of model heparin-binding proteins to immobilized monosaccharides and clusters of carbohydrates. They evaluated the effect of ligand density and ligand clustering by measuring the binding of a binding domain of vWf to the SPR chip. They found that clusters of oligosaccharides were more effective ligands than monosaccharides, even when comparable (per carbohydrate) surface loadings were present.[123] In addition to using SPR to compare loading densities, Fieschi and co-workers used mannose-functionalized Boltorn hyperbranched dendrimers to inhibit the binding of DC-SIGN to a gp120-functionalized surface.[124] Kiessling and co-workers used SPR with mannose functionalized surfaces to determine the concentrations of neoglycopolymers that inhibited the binding of Concanavalin A (Con A) to the surface. The neoglycopolymers varied in length, and the most potent compounds were those that could bind to multiple Con A binding sites as well as cross-link the Con A lectins.[125] Riguera and co-workers compared SPR results for mannose-functionalized dendrimers binding to surface bound Con A with the inhibition of Con A binding to mannose-functionalized gold surfaces.[126] Schengrund and co-workers used SPR to study the binding of HIV-1 gp120 to potential glycodendrimer inhibitors. They found that binding affinities of the glycodendrimers with rgp120 (as measured by SPR) and the ability of the glycodendrimers to inhibit HIV infectivity (as measured by viral inhibition assays) correlated well for sulfated glycodendrimers.[127] Imberty, Matthews, Vidal and co-workers used SPR inhibition assays to test the potential of galactose-functionalized calyx[4]arenes to block adhesion of PA-IL to galactosylated surfaces.[128] These examples indicate that applying SPR technology to the investigation of protein-carbohydrate interactions shows great promise.

Because many lectins of current interest for biological processes are not surface bound, robust biosensors for the measurement of multivalent binding interactions in the solution phase are needed. Dithiol compounds are an ideal system with which to build a self-assembled monolayer on a gold surface. The dithiols form a bivalent attachment to the gold that is robust enough to undergo many inhibition-binding measurements. Here, the effectiveness of different generations of mannose-functionalized dendrimers to inhibit binding of Con A to a mannose-functionalized surface through SPR is described.

Self-Assembled Monolayers

Self-Assembled monolayers (SAMs) are well-ordered arrays which allow for fundamental studies of interactions occurring at surfaces. Because of their well-defined organization and ease of functionalization, they are ideal model systems for surface-based interactions. One of the most widely used systems to form SAMs is the gold-alkylthiolate monolayer first produced in 1983.[129] These monolayers are produced by immersion of a gold surface in an ethanolic solution of the thiol, and are stable for several months.[130] Mixed monolayers are produced if the ethanolic solution contains more than one thiol type. Since its commercial induction, surface plasmon resonance (SPR) spectroscopy has become an important technique for measuring reactions occurring on a monolayer surface.

Reaction kinetics on a surface can vary greatly from reactions occurring in solution; this is due to the rate of the reaction changing as the surface becomes more modified, and the close proximity of the reactive sites and alkyl tethers.[131] Generally, sterics play a large part in the reactivity of the surface monolayers. S_N2 reactions are considerably slowed on planar SAMs, even for small nucleophiles.[132] Despite this difficulty, SAMs retain a number of applications in organic chemistry. Catalytic reactions requiring a defined presentation, such as those found in ring-opening metatheses, benefit from the immobilization and afford similar yields to solution-based reactions.[133, 134]

Molecular recognition is an area where SAMs see important applications. Using techniques such as surface plasmon resonance allows for high throughput screening of molecules with a tethered substrate. Non-specific interactions are greatly reduced, as non-specific binders can be simply washed over the surface, while specific binders recognize the tethered substrate. Competitive and inhibition studies can be monitored in real-time and with high precision, and are less subject to possible interfering molecules present in solution. In cases where the binding is reversible, using a SAM offers better reproducibility of the experiment, as the same SAM can be used for multiple experiments.

A SAM of dithiols on an unmodified gold surface was created.[135, 136] Alkanemonothiols are known to spontaneously desorb from gold surfaces under ambient conditions, resulting in patchy surfaces with significant areas of metal exposed.[137] The stability of the surface can be significantly improved through the use of multivalent dithiol compounds.[138, 139] The carboxyl-terminated dithiol compounds were coupled to amine-functionalized mannose derivatives through amide bond formation. This allowed for a stable mannose surface to be presented on the gold chip, which could be reused for multiple assays with negligible degradation over time (Figure 3.2).

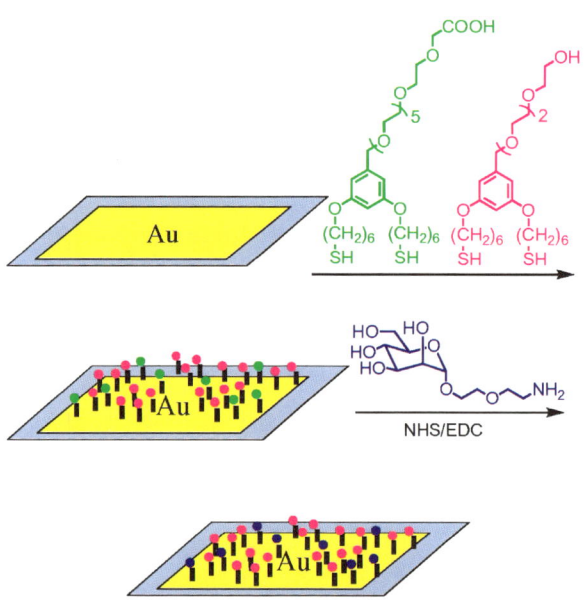

Figure 3.2. A schematic representation of mannose functionalization of a gold surface.

Although a commercially available gold surface coated with dextran, a polymer composed of glucose residues, is commonly used in SPR assays, competition experiments involving mannose-functionalized dendrimers and Con A proved to be problematic with this chip. Although Con A has a 3-4 fold higher affinity for mannose than for glucose[140], the change in instrument response elicited by the addition of mannose-functionalized dendrimer was small compared to the binding of Con A to the dextran matrix in the absence of dendrimer. The minimal response change likely occurred because of the large excess of glucose residues on the dextran-coated gold surface relative to the number of mannose residues on the glycodendrimers. Furthermore, using a dextran surface with large multivalent glycodendrimers may induce sugar-sugar binding events, complicating the analysis of the specifically targeted binding events. The mannose-functionalized monolayers used for this research were extremely robust because of the use of dithiols and were used repeatedly for at least six months without detectable degradation.

Surface Plasmon Resonance

Concanavalin A is tetrameric plant lectin that is able to specifically bind to mannose residues.[31] Serial dilutions of Con A were injected, and the SPR response was monitored to determine the affinity of Con A for the mannose-functionalized surface. The K_d for the binding of Con A to the mannose-functionalized surface was estimated at 78 nM (Figure 3.3).

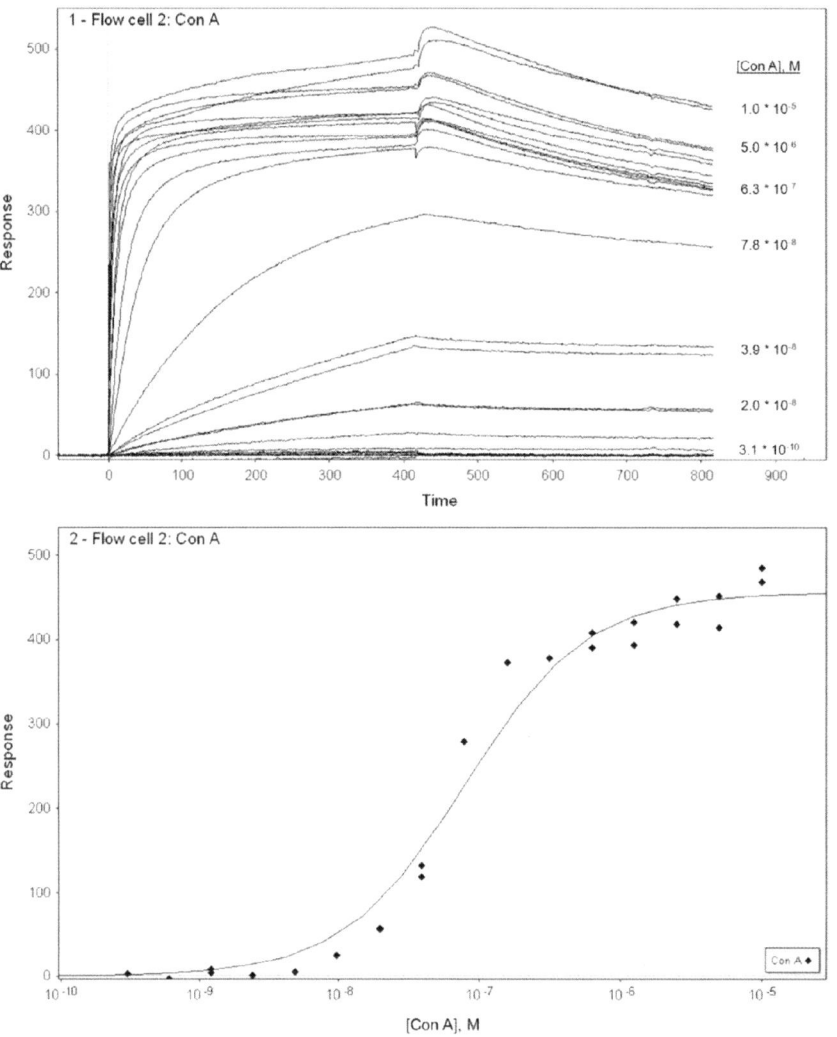

Figure 3.3. Top: Doubly-referenced sensorgram of Con A binding to the mannose-functionalized surface. Bottom: Affinity profile of Con A to the mannose-functionalized surface, fit using Scrubber 2.

The glycodendrimers that were used in this study are shown in Figure 3.4a, and the biosensor strategy is shown in Figure 3.4b. The values for the number of mannosides were determined from M_W (MALDI-TOF MS)[52, 105] and are the average number of mannosides per dendrimer. For a discussion of the homogeneity of the dendrimers, see reference 141. The affinity of the glycodendrimers for Con A was evaluated by studying their ability to inhibit binding of Con A to the mannose-functionalized surface. Serial dilutions of the inhibitor were pre-mixed with a constant amount of Con A, and the resulting equilibrium responses were used to determine IC_{50} values (Figures 3.5 and 3.6). A key point is that glycodendrimer/lectin binding occurs in solution before being passed over the monolayer surface, better mimicking protein-carbohydrate interactions.

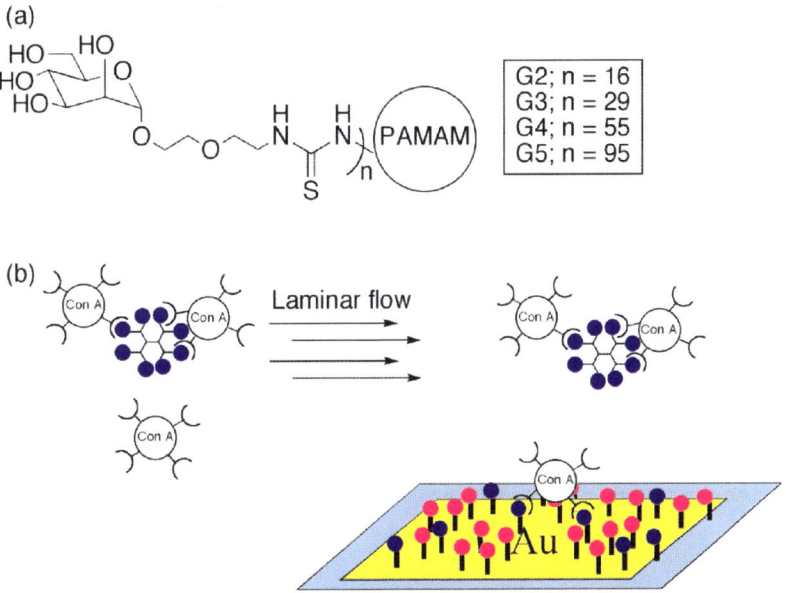

Figure 3.4. (a) Mannose-functionalized poly(amidoamine) (PAMAM) dendrimers. (b) A schematic representation of the inhibition binding experiment.

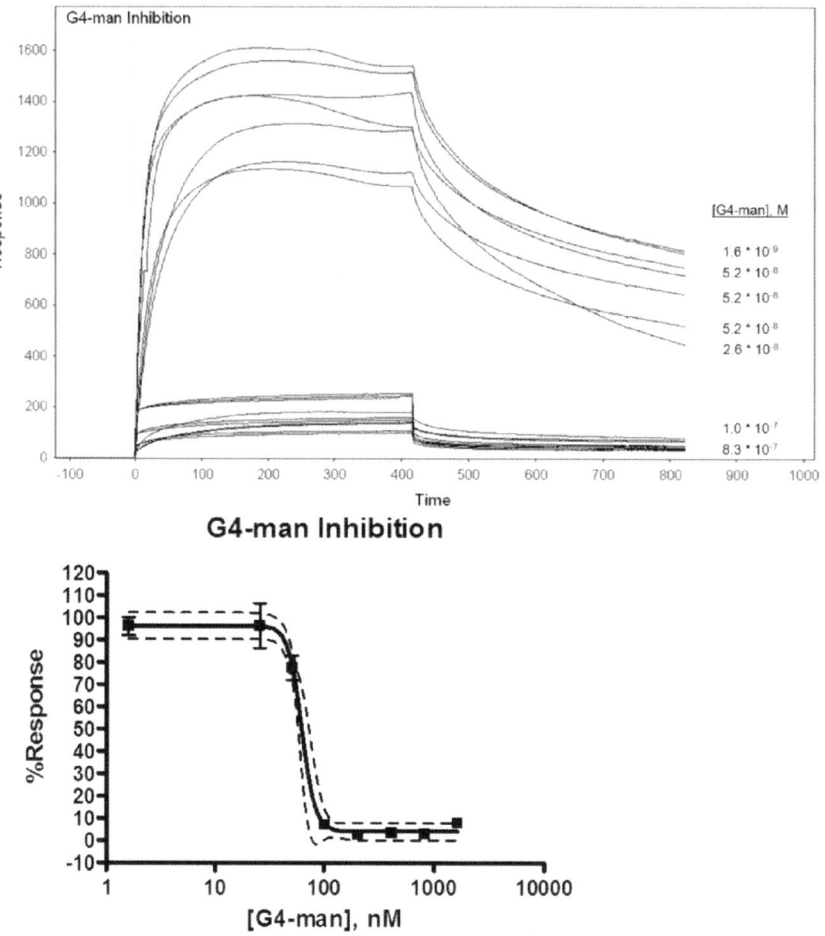

Figure 3.5. Inhibition of 2 μM Con A injected over the mannose-functionalized gold surface, normalized to 100% response from Con A without addition of inhibitor. Top: Sensorgram of Con A binding to the mannose-functionalized surface, inhibited by mannose-functionalized G4 dendrimer. Bottom: Inhibition by mannose-functionalized G4 dendrimer. Dotted lines show 95% confidence levels, fit using GraphPad Prism 4.

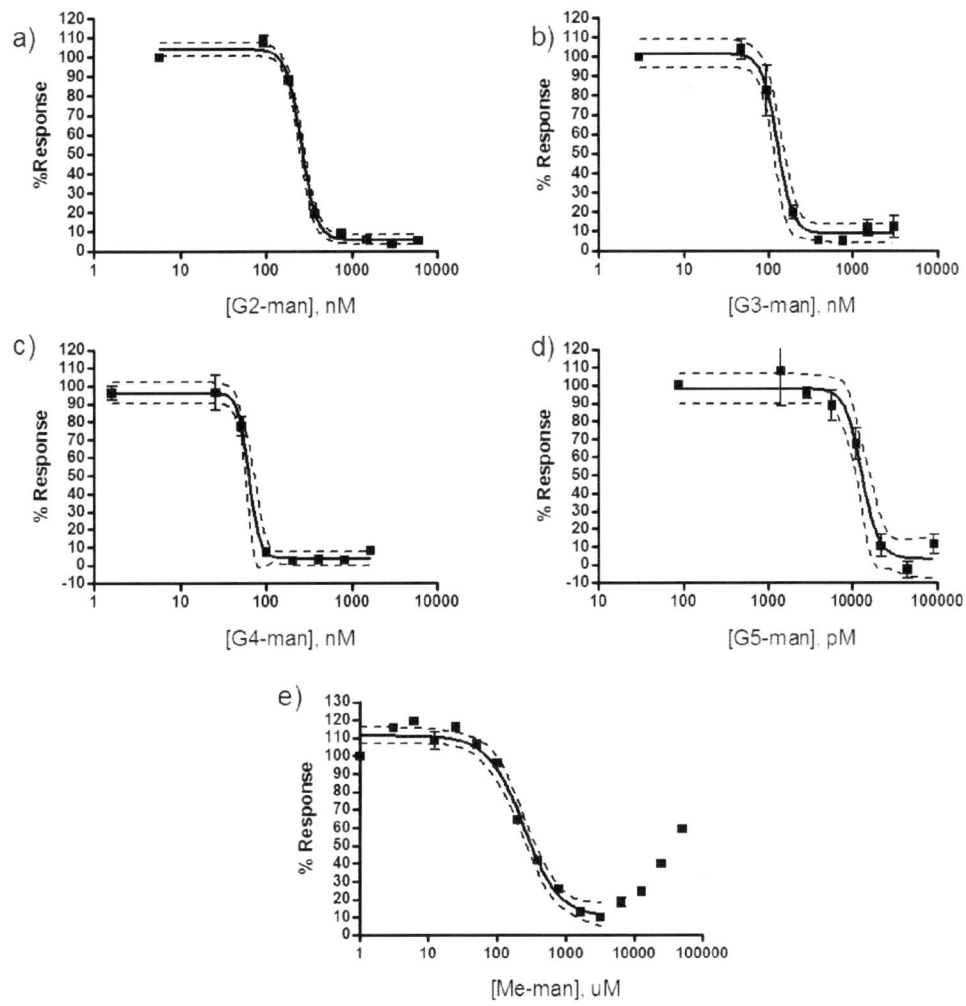

Figure 3.6. Inhibition by mannose-functionalized PAMAM dendrimers of 2 μM Con A. Dotted lines show 95% confidence levels, fit using GraphPad Prism 4. (a) Mannose-functionalized G(2)-PAMAM, (b) mannose-functionalized G(3)-PAMAM, (c) mannose-functionalized G(4)-PAMAM, (d) mannose-functionalized G(5)-PAMAM, (e) Methyl-mannoside.

Whereas a marked increase in activity with increasing glycodendrimer generation was previously observed by hemagglutination assays[52, 105], the IC_{50} values obtained by

SPR suggest that the ability to inhibit Con A binding to the surface is largely independent of dendrimer generation. While the glycodendrimers offer a definite advantage over methyl-mannose, higher generation dendrimers exhibit similar abilities to inhibit binding when compared to lower generations. Binding data is summarized in Table 3.1.

Table 3.1. Comparison of Hemagglutination Inhibition Assays with SPR results for mannose-functionalized dendrimers with Concanavalin A.

Hemagglutination Inhibition Assay Results[a,b]		SPR Inhibition Assay Results[a]	
Dendrimer Generation (number of sugars)	Relative activity per mannose (MIC)[c]	IC_{50} values (per dendrimer)[d]	IC_{50} values (per mannoside)[d]
Methyl mannose (1)	1 (6400 μM)	240 μM	240 μM
G2 (16)	~1 (9700 μM)	260 nM	4.2 μM
G3 (29)	20 (280 μM)	130 nM	3.8 μM
G4 (55)	200 (30 μM)	63 nM	3.5 μM
G5 (95)	300 (20 μM)	13 nM	1.2 μM

[a] Each reported value represents at least three assays.
[b] Values are taken from references 105 (G2) and 52 (G3–G5).
[c] Minimum Inhibitory Concentration per mannoside for 0.7 μM Con A.
[d] Inhibition of 2 μM Con A.

SPR assays in which Con A was bound to a mannose-functionalized gold surface indicated that the K_d for the binding of Con A to the mannose-functionalized surface was 78 nM. This value indicates a significantly stronger interaction compared to the reported K_a of 7.6×10^3 M^{-1} (K_d of 130 μM) obtained for monovalent methyl-mannose from microcalorimetry.[142] This disparity almost certainly occurs because Con A is able to bind multiple surface sugars, increasing the overall affinity of Con A for the self-assembled monolayer.[41]

The gold surface used was functionalized with a rather high percentage (25%) of a carboxyl-terminated dithiol compound that was subsequently coupled to a tethered mannose derivative. This places the mannosides close enough to undergo bivalent binding with the tetrameric Con A, preventing an accurate determination of monovalent association and dissociation rate constants but allowing equilibrium constants of highly multivalent systems to be determined. The affinity profile for Con A, shown in Figure 3.3, does not significantly fit a 1:1 kinetic binding model (Figure 3.7). Heterogeneous binding, which is a more effective mimic of relevant biological surfaces is suggested and allows us to study the ability of the dendrimers to inhibit multivalent binding modes to a surface. Because comparisons are of the inhibitory propensity of different generations of glycodendrimers for Con A binding to the monolayer, determination of the exact degree of mannose incorporation into the monolayer was unnecessary. Although other researchers have used attenuated total reflection FT-IR spectroscopy to determine the degree of SPR chip functionalization[123], this was not required for our studies because the same chip was used in all of the experiments, eliminating variability due to different chips.

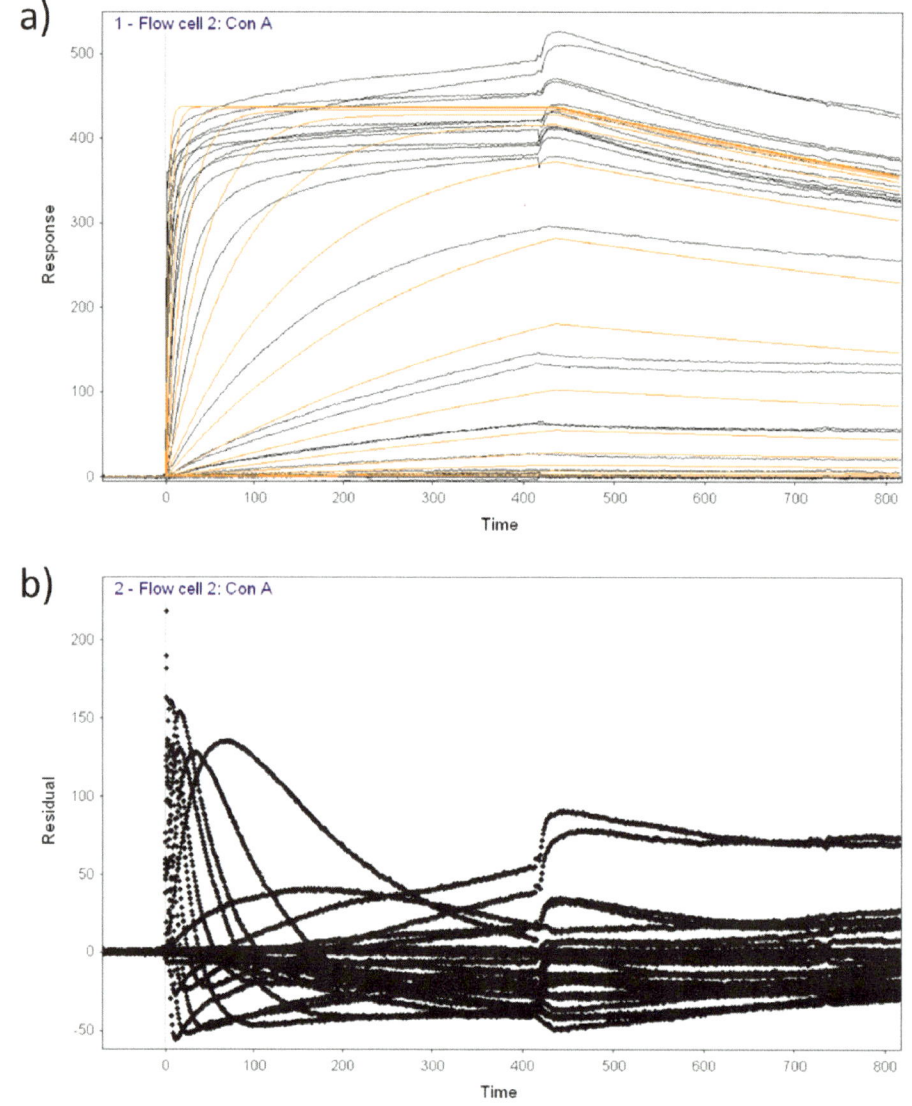

Figure 3.7. a) Attempted fit of Con A to a kinetic 1:1 binding model. Projected fit is shown in red. b) Residual difference between the data set and the 1:1 model fit.

While the results obtained by SPR show that the glycodendrimers exhibit a definite advantage over the monovalent sugar, the increased activity observed with higher generations of dendrimers in hemagglutination inhibition assays with Con A is not evident in SPR assays. In fact, the mannose-functionalized dendrimers showed only small affinity differences (on a per mannose basis), compared to remarkable activity increases observed in hemagglutination assays.

In contrast to the hemagglutination assay, SPR offers a method to more directly measure the interactions occurring between Con A and the sugar. In the hemagglutination inhibition assay, red blood cells are added as sources of surface sugars, which are bound by Con A. Inhibition of this interaction by glycodendrimers provides an entry-level comparison of our system to known systems, although the assay is not able to provide information regarding binding affinity. Rather, hemagglutination indicates the ability of a glycodendrimer to drive aggregation processes, and not the strength of protein-carbohydrate binding[44, 143]. As a result, SPR provides a more accurate measurement of protein-carbohydrate affinity independent of aggregation and precipitation events.

Although the mannose-functionalized surface is successfully bound by Con A, additional binding events can be observed in the inhibition sensorgrams. Non-specific binding arising from sugar-sugar interactions is prominently displayed at high concentrations of methyl-mannose (Figure 3.6e). At sufficiently high concentrations of methyl-mannose, Con A is prevented from binding to the surface. Even higher concentrations of methyl-mannose, however, afford an increased response through nonspecific sugar-sugar interactions. This suggests significant sugar-sugar interactions; a pre-organization step between sugars may be a factor in cellular recognition, as opposed to relying solely on the specificity of a protein for a sugar residue.[144, 145]

Because of the heterogeneous nature of the binding events, kinetic data cannot be accurately determined from this experiment. Several binding events are occurring, including multivalent binding motifs, which make it difficult to de-convolute individual binding events from the overall system. However, information can be obtained from equilibrium data, particularly about the efficiency of the tested inhibitors. Examining the response as a function of added glycodendrimer reveals the efficiency at which the glycodendrimers are able to inhibit Con A from binding to the surface of the self-assembled monolayer.

Of particular interest is the increased steepness of the curve when the inhibition data is fit to the four-parameter logistic equation (Figure 3.5, bottom; Table 3.2). Whereas inhibition of the Con A–surface sugar interaction by methyl-mannose yields a rather shallow slope, inhibition by glycodendrimers yield markedly steeper slopes. By contrast, inhibition by galactose-functionalized dendrimers could not be fit to the equation and yielded similar responses throughout the dilution series (Figure 3.8). This suggests that addition of mannose-functionalized dendrimer to the system induces a pronounced effect on Con A binding, and that responses from the assay are more sensitive to changes in dendrimer concentration than in methyl-mannose concentration. As the amount of glycodendrimer that is added is increased, Con A sequestration (such that less Con A is available to be bound to the self-assembled monolayer) is enhanced. This is indicative of the dendrimers' ability to effectively bind Con A and to isolate bound Con A from a sugar-functionalized surface, as expressed by IC_{50} values (Table 3.1, Table 3.2).

Table 3.2. Curve fit parameters for inhibition experiments determined by GraphPad Prism 4.

Sigmoidal dose-response (variable slope)	G2-man	G3-man	G4-man	G5-man	me-man
Best-fit values					
BOTTOM	5.872	8.977	3.894	3.349	10.14
TOP	104.3	101.7	96.24	98.31	111.3
LOGEC50	2.416	2.106	1.799	4.117	2.386
HILLSLOPE	-5.011	-4.976	-6.805	-4.276	-1.664
EC50	260.8	127.6	63	13093	243.1
Std. Error					
BOTTOM	1.215	2.323	1.849	5.362	3.823
TOP	1.674	3.491	2.796	4.019	2.144
LOGEC50	0.01444	0.03139	0.02576	0.04111	0.04141
HILLSLOPE	0.463	1.065	1.78	1.614	0.2487
95% Confidence Intervals					
BOTTOM	3.318 to 8.425	4.053 to 13.90	-0.04548 to 7.833	-7.873 to 14.57	2.076 to 18.21
TOP	100.8 to 107.8	94.30 to 109.1	90.28 to 102.2	89.90 to 106.7	106.8 to 115.9
LOGEC50	2.386 to 2.447	2.039 to 2.173	1.744 to 1.854	4.031 to 4.203	2.298 to 2.473
HILLSLOPE	-5.984 to -4.038	-7.233 to -2.719	-10.60 to -3.012	-7.654 to -0.8975	-2.188 to -1.139
EC50	243.2 to 279.6	109.5 to 148.8	55.52 to 71.48	10740 to 15962	198.8 to 297.3
Goodness of Fit					
Degrees of Freedom	18	16	15	19	17
R^2	0.9938	0.9738	0.9856	0.9334	0.984
Absolute Sum of Squares	280.2	911.3	457.8	2967	578.1
Sy.x	3.945	7.547	5.524	12.5	5.832

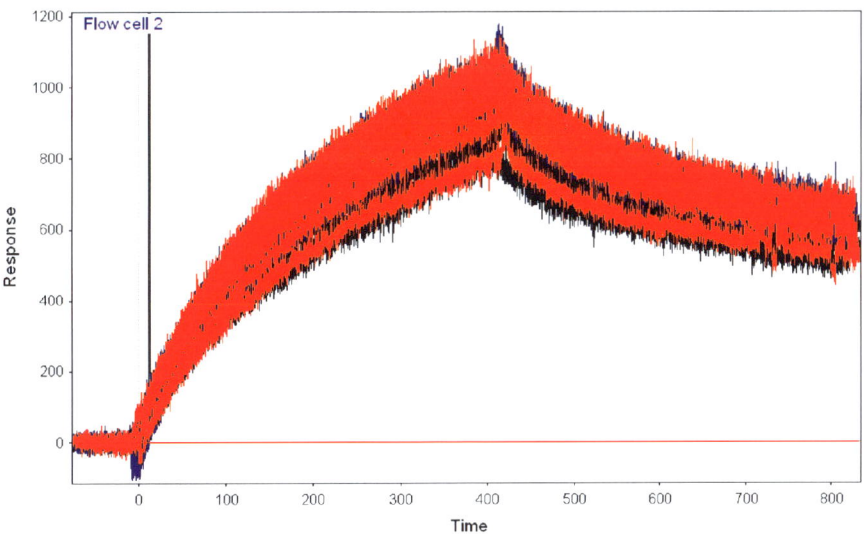

Figure 3.8. Inhibition by serial dilutions of galactose-functionalized G4 dendrimer, represented by three overlaying dilution series in Scrubber 2.

The Four Parameter Logistic Equation

The four parameter logistic equation is defined by four parameters: the baseline response (Bottom), the maximum response (Top), the slope (Hill Slope), and the concentration of the compound which provokes a response halfway between the baseline and the maximum (EC_{50}) (Equation 3.1). For inhibition studies, the EC_{50} is usually called the IC_{50}, or the concentration which *inhibits* 50% of the response, as defined by the 100% response (Top) and 0% response (Bottom). Depending on how the data is normalized, the IC_{50} may not be the concentration that is derived from a response of Y = 50; the IC_{50} is merely the concentration that evokes a response halfway between the maximum and the baseline.[146]

$$response = Bottom + \frac{(Top - Bottom)}{1 + 10^{(LogEC50 - Log[Ligand]) * HillSlope}}$$

Equation 3.1. The four parameter logistic equation.

Some binding curves are steeper or shallower than standard curves, expressed by a slope called the Hill slope. Whereas a standard curve has a Hill slope of 1, a steeper curve has a larger slope and a shallower curve has a smaller slope. For inhibition studies, this slope is negative, with steeper slopes having more negative values. In most cases, the Hill slope cannot be interpreted to yield significant chemical information, other that a Hill slope significantly different from 1 does not follow the typical law of mass action for the interaction, meaning that simple kinetics do not apply.[146] It is difficult to interpret the meaning of the derived IC_{50} in terms of standard mechanistic mass-action models.

Significantly steeper slopes can be an indication of positive cooperativity, for example, if the binding sites for the system are clustered together. In the case of dendrimer inhibition, a significantly steep Hill slope may be due to the fact that added dendrimer to the SPR system creates more effective cross-links than a lower concentration of dendrimer is able to. For this reason, the efficacy of the dendrimers to sequester Con A from the SPR surface is expressed as IC_{50} values as a measure of inhibitor potency.

Conclusions

Highly robust mannose-functionalized SAMs for inhibition binding studies using SPR were formed on gold surfaces using dithiols. SPR experiments with these dithiol SAMs demonstrated that glycodendrimers efficiently inhibit protein-carbohydrate interactions; the potencies reported here for glycodendrimers are considerably higher than those obtained for monovalent mannose (on a per mannose basis). However, the generation of the glycodendrimer had less of an impact in these SPR studies than was observed in previously reported hemagglutination inhibition assays.[52, 105] That the amount of inhibitor needed to achieve complete inhibition of red blood cell aggregation is decidedly different from the IC_{50} values reported here for SPR assays is readily rationalized because of fundamental differences in what the assays measure. In a hemagglutination inhibition assay, glycodendrimers probably attenuate aggregation events between Con A and red blood cells by efficiently sequestering the Con A lectins from the red blood cells. In assays that do not rely on disruption of aggregation events, dendrimers exhibit solely sugar-protein interactions, with less pronounced increases in affinity due to glycoside clustering effects.[44, 142] The SPR competition assays reported here reaffirm that affinity enhancements for multivalent ligands are present. These SPR competition assays are an effective tool for studying solution phase binding events.

Glycodendrimers are very potent frameworks for mediation of protein-carbohydrate interactions. They are capable of both forming and disrupting extensive cross-links. When designing therapeutic agents for human health, our results suggest that different multimeric glycosides can exhibit similar activities for inhibiting protein-carbohydrate interactions, but decidedly different activities for interacting with large existing complexes are likely.

Experimentals

General Methods. Mannose-functionalized dendrimers were synthesized as previously described in 105. Solutions and running buffers were degassed before use in SPR experiments. Unless otherwise noted, 10 mM PBS, pH 7.4 was used as running buffer for all SPR experiments.

SPR experiments were performed using a Biacore 1000 upgrade equipped with a Sensor Chip Au (Biacore AB, Uppsala, Sweden). The instrument temperature was set to 25 °C. Dialkanethiols (SPT-0013 and SPT-0014) were purchased from SensoPath Technologies (Bozeman, MT) for functionalization of the gold surface. Concanavalin A was purchased from CalBioChem (Darmstadt, Germany). All other chemical reagents were purchased from Fisher Scientific (Waltham, MA).

Preparation of 2-(2-aminoethoxy)ethyl-α-D-mannopyranoside. 2-(2-azidoethoxy)ethyl-2,3,4,6-tetraacetyl-α-D-mannopyranoside (330 mg, 0.74 mmoles) was dissolved in 10 mL of MeOH and reduced over Pd/C and an excess of H_2. The catalyst was removed and the resulting mixture was filtered through a 0.45 μm filter. Removal of the solvent *in vacuo* yielded 217 mg (79% yield) of 2-(2-aminoethoxy)ethyl-2,3,4,6-tetraacetyl-α-D-mannopyranoside, which was used directly.

2-(2-aminoethoxy)ethyl-2,3,4,6-tetraacetyl-α-D-mannopyranoside (217 mg) was dissolved in 10 mL of 1:1 MeOH:H_2O. The acetyl groups were then removed with 1M NaOMe in MeOH (522 μL, 522 μmoles), stirring overnight. The solvent was removed under vacuum. The resulting sugar and acetic acid salts were dissolved in Millipore water and were used without further purification. ^{1}HNMR (d$_6$-DMSO, 500 MHz, solvent peaks were visible but not recorded) δ 3.2-3.7 (m, 21 H). ^{13}CNMR (d$_6$-DMSO, 125 MHz, solvent peaks were visible but not recorded) δ 30.1, 40.5, 61.7, 66.1, 67.5, 69.8, 70.7, 71.4, 74.4, 100.4. These results agree well with previously published data for this compound, which was synthesized via a slightly different route and for which NMR data is reported in D$_2$O.[147]

Preparation of Carboxylated Gold Surface. 25% of a -COOH terminated aromatic dialkanethiol (SPT-0014, SensoPath Technologies) was mixed with 75% of a -OH terminated aromatic dialkanedithiol (SPT-0013, SensoPath Technologies) to provide a 1 mM dithiol solution in Ethanol. This solution was diluted to 10% (v/v) in Millipore water and was injected over a single flowcell area (flowcell 2) of a new gold chip several times at 1 µL/min for 60 minutes using Millipore water as running solution, until no appreciable signal increase was observed. This procedure was repeated using 100% -OH terminated aromatic dialkanethiol (SPT-0013) over flowcell 1 on the same gold chip for a reference cell.

Preparation of Mannose-functionalized Gold Surface. The carboxylated gold surface was activated by injection of a mixture of *N*-ethyl-*N'*-(diethylaminopropyl)-carbodiimide (EDC) and *N*-hydroxysuccinamide (NHS) (30 µL, 200 mM EDC, 50 mM NHS, 5 µL/min) dispensed by the Biacore instrument. 2-(2-aminoethoxy)ethyl-α-D-mannopyranoside (10 mg/mL in 10 mM PBS, pH 7.2, includes acetic acid salts from the deprotection step) was injected into flowcell 2 (12 min contact time) and allowed to react.

This procedure was repeated three times to allow for maximum functionalization. Unreacted activated carboxyl groups were given extensive time to hydrolyze under running buffer; remaining activated carboxyl groups were then capped by injection of ethanolamine (20 μL, 1mM pH 8.5).

Surface Plasmon Resonance. Concanavalin A (Con A) (100 μL in 10 mM PBS, pH 7.4) was injected over the mannose-functionalized surface at 10 μL/min, allowing 300 s for dissociation, followed by regeneration (5 μL, 10 mg/mL methyl-mannose in 0.1 M HCl). The titration range covered 10 μM to 31 pM by 2-fold dilutions. K_a was estimated using Scrubber 2.0.

For competition experiments with inhibitors, Con A (2 μM monomer) was mixed with an equal volume of inhibitor and the solution was allowed to equilibrate for at least 1 h. The mixture (70 μL) was injected over the surface at 10 μL/min, allowing 300 s for dissociation, followed by regeneration (5 μL, 10 mg/mL methyl-mannose in 0.1 M HCl). The glycodendrimers were tested in triplicate, covering a range of 50 μg/mL to 39 ng/mL by 2-fold dilutions. For competition experiments with methyl-mannose, the assay covered a range of 10 mg/mL to 305 ng/mL by 2-fold serial dilutions. Equilibrium responses were determined using Scrubber 2.0, then analyzed in GraphPad Prism (version 4.0) using the four-parameter logistic equation to determine IC_{50} values.

Binding responses were obtained using the aforementioned conditions across a mannose-functionalized surface. The binding responses were referenced by subtracting the response generated from identical injection conditions over a flowcell containing only

–OH terminated dialkanethiols (flowcell 1), and double-referenced by subtracting a response from buffer injections.

CHAPTER 4

TRISMANNOSE-FUNCTIONALIZED DENDRIMERS TO INVESTIGATE CLUSTERING EFFECTS BY ELISA

Introduction

The mammalian cell surface is decorated with a dense layer of carbohydrates, many of which are cell-specific and participate in a wide variety of biological processes.[148, 149, 150] Abnormal glycosylation patterns have been associated with the infection and proliferation of many diseases, and are particularly known to play a crucial role in promoting or inhibiting the metastasis of cancer.[151, 152, 153] Because glycosylation is one of the most common modifications made to cells, there is a need to examine the architecture displayed on cell surfaces with glyco-mimetic compounds.

Patterning and Clustering

Several research efforts focused on exploring surface sugar patterning have made significant contributions to improved understanding of surface architecture and binding motifs.[154, 155-171] Lindhorst et al. have created numerous multivalent glycoside clusters and glycodendrons, investigating the effects of spacer compounds and the cluster effect of several mannosides.[156, 157, 158, 159] Ru(bipy)$_3$-based glycodendrimers and fluorescently labeled glycoclusters have been synthesized to investigate binding.[160, 161, 162] Glycoclusters provide a way of targeting carbohydrate binding proteins with a higher affinity than monomeric units can provide. This increase in affinity has been attributed to a "cluster glycoside effect", which provides an "affinity enhancement over and beyond what would be expected from the concentration increase of the determinant sugar in a multivalent ligand".[40, 44, 163]

Cluster glycosides offer a convergent synthetic route to create regular, reproducible patterns on small molecules. Click chemistry is an efficient means to couple glycoside clusters onto a myriad of scaffolds without major changes to the synthetic route.[164, 165] Glycosides have been "clicked" onto a variety of surfaces, ranging from nanoparticles in biological assays to gold surfaces for use with surface plasmon resonance.[166, 167, 168] Roy et al. have used this method to create clustered glycodendrons based on several scaffolds.[169, 170, 171] A dendritic architecture employed was based on tris(hydroxymethyl)aminomethane, a component in the common TRIS buffer. This allowed for an expedient route to the creation of larger cluster-based glycodendrimers, the method employed in this study.

Dendrimers provide a regular, highly customizable scaffold ideal for probing multivalent and scaffolding effects (Figure 4.1). Previous studies have shown evidence that surface features such as shape complementarity play an important role in binding, as well as the method ligands are displayed (cf. previous chapters). The type and number of carbohydrates displayed on a surface help control the strength of binding, and can be appreciably controlled using a regular scaffold such as a dendrimer, revealing binding and inhibition efficacies dependent on size and functionalization.[52, 53, 105] Display of monomeric carbohydrates offer a great degree of customizability. This study describes the synthesis of multiple carbohydrate clusters affixed to higher generation PAMAM dendrimers and their inhibition efficacy in ELISA assays to investigate the effect of closely spaced surface patterning on large scaffolds.

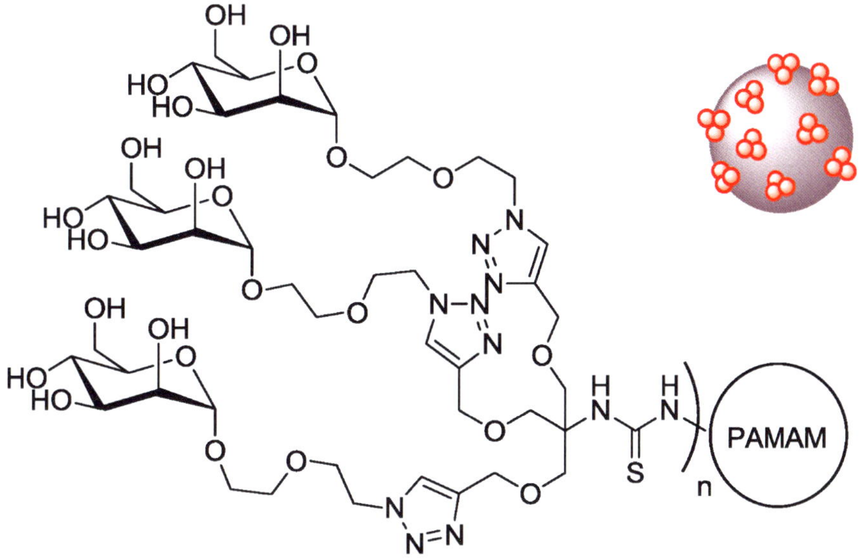

Figure 4.1. Trismannose-cluster-functionalized PAMAM dendrimer. Top: representative model.

Synthesis of Trismannose-Functionalized Dendrimers

Sugar clusters were synthesized using "click chemistry", a straightforward, highly efficient process which yields no byproducts and can be performed in water.[164, 165] The trismannose-functionalized dendrimers were prepared as previously described by Joel Morgan.[61] Briefly, synthesis of the sugar cluster begins with the readily available tris(hydroxymethyl)aminomethane (TRIS) to create a tris-propargyl ether used as the alkyne component for the copper mediated addition of three equivalents of an azido-functionalized mannose.

PAMAM dendrimers were each partially functionalized with spacer compound **7**, 2-(2-isothiocyanatoethoxyethanol), followed by functionalization with sugar cluster **8** (Figure 4.2, Scheme 4.1).[61, 105] After purification, deprotected heterogeneously functionalized glycodendrimers **17-20** were obtained.[61, 105] As a control, unclustered mannose functionalized dendrimers were synthesized as previously described (Figure 4.3).[52, 53, 105] These glycodendrimers were assembled by taking advantage of the isothiocyanate functionality as described above, but had only a single mannose residue per dendrimer endgroup. This allows for evaluation of the effects of clustering sugar residues in close proximity versus a random functionalization of the dendrimer surface.

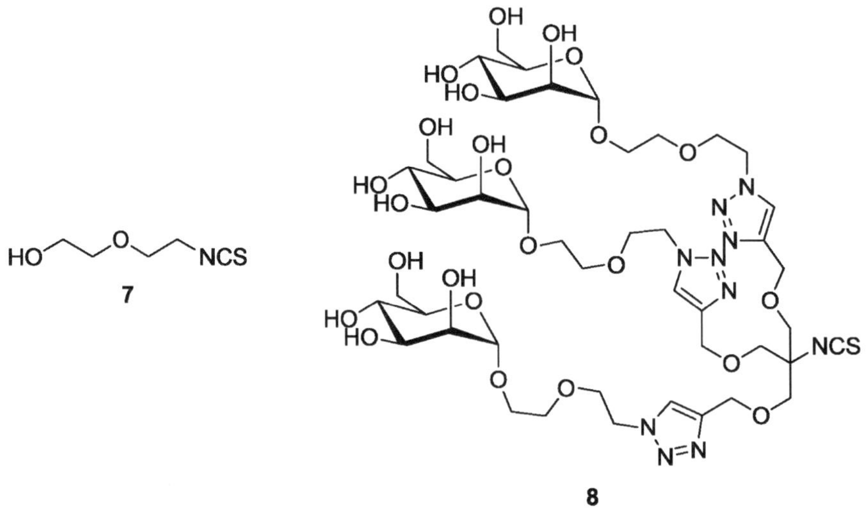

Figure 4.2. Spacer compound **7**, sugar cluster **8**.

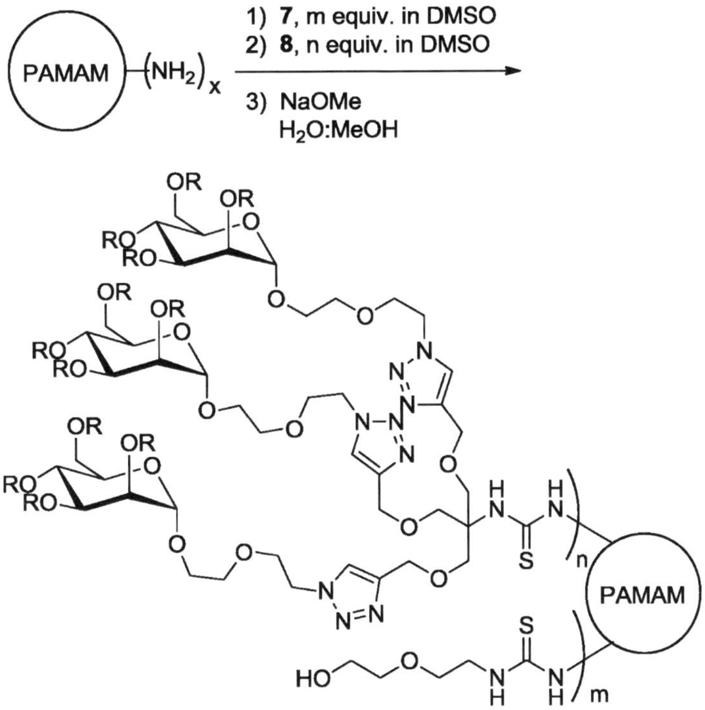

Scheme 4.1. Synthesis of deacetylated heterogeneously functionalized tris-mannose cluster and ethoxyethanol PAMAM dendrimers. (cf. reference 61)

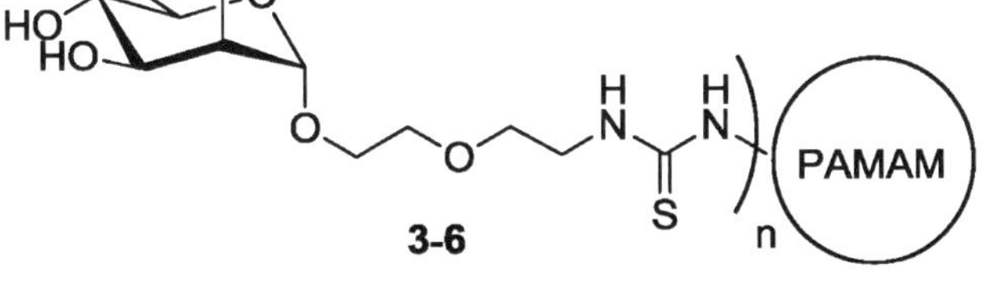

Step 1:	Step 2: R = Ac	Step 3: R = H
G(3) **9a-9d**	G(3) **13a-13d**	G(3) **17a-17d**
G(4) **10a-10g**	G(4) **14a-14g**	G(4) **18a-18g**
G(5) **11a-11f**	G(5) **15a-15f**	G(5) **19a-19f**
G(6) **12a-12e**	G(6) **16a-16e**	G(6) **20a-20e**

3 G(3); n = 29
4 G(4); n = 55
5 G(5); n = 95
6 G(6); n = 178

Figure 4.3. Mannose functionalized dendrimers **3-6**.

SDS-PAGE of Glycosylated Dendrimers

The molecular weight of the glycosylated dendrimers was determined by MALDI-TOF. However, several dendrimers in the series did not yield expected molecular weights, and did not give expected results in preliminary ELISA Assays. Because MALDI-TOF MS of trismannose-functionalized dendrimers is difficult due to solubility issues, and to test alternative methods of determining the molecular weight of glycodendrimers, sodium dodecyl sulfate polyacrylamide gel electrophoresis (SDS-PAGE) was used as a test to determine whether the compounds had degraded.

SDS-PAGE is a technique used to separate molecules based on their electrophoretic mobility. The SDS eliminates differences based on electronic charge, as the molecules become saturated with the small molecules and confers a large negative charge to them. As a result, separation should occur solely based on the size of the molecules.[172] Using a modification of Periodic Acid-Schiff methods[173, 174], the carbohydrates on dendrimers can be visualized on the gel as magenta bands (Figure 4.4).

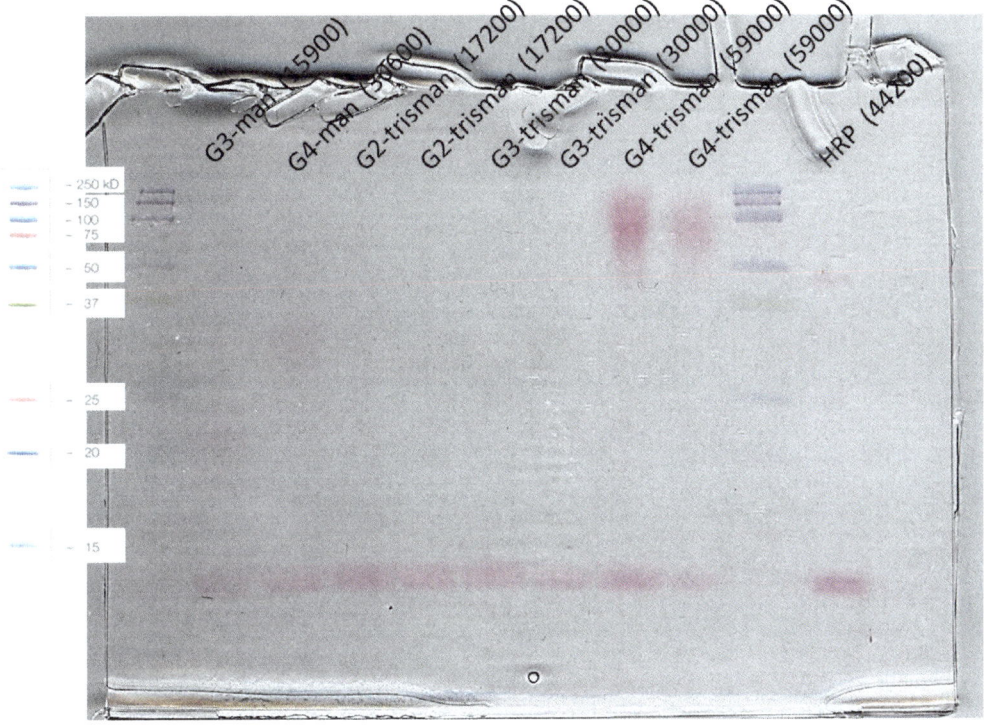

Figure 4.4. SDS-PAGE of glycodendrimers visualized by a modified Periodic Acid-Schiff method on a 15% Gel. Expected molecular weights are shown in parentheses. (HRP = Horseradish Peroxidase)

The visualized bands appear significantly more diffuse than typical protein standards, perhaps because of the heterogeneity inherent in functionalized glycodendrimers. Furthermore, due to the highly branched nature of the dendrimer, the bands appear at a higher molecular weight than expected. For compounds which did not give expected molecular weights using MALDI-TOF, SDS-PAGE confirmed that the compounds had decomposed; there were no sugars present on a dendrimer scaffold. The control glycodendrimers and glycoprotein however, yielded a result congruent with the expected molecular mass. For this reason, further experiments were only continued with glycodendrimers that presented an expected molecular mass, while compounds that did not show a correct molecular mass were assumed decomposed and were not experimented upon.

Enzyme-Linked Immunosorbent Assays

To gain insight into the mechanism of protein-carbohydrate interactions, the synthesized glycodendrimers' inhibition efficacy were analyzed as ligands in ELISA inhibition experiments. For these experiments, Concanavalin A (Con A) was used, as it is an inexpensive, robust, homotetrameric lectin known to selectively bind mannose and glucose residues.[31, 175] Con A has also been often used in our lab as a tool to study multivalent interactions and has been used to characterize the binding behavior of previously synthesized glycodendrimers.[52, 53, 105] Briefly, 96-well plates were covered with RNase B, a high mannose glycoprotein. Bovine Serum Albumin (BSA) was then added to block any part of the surface not occupied by RNase B. Pre-incubated mixtures of glycodendrimer and Con A were added to the plate and left to reach equilibrium. After washing away any unbound Con A, enzyme-linked anti-Con A was added. 4-nitrophenyl phosphine was added as a substrate for the enzyme; the amount of Con A attached to the surface could then be determined by the absorbance at 410nm (Figure 4.5).

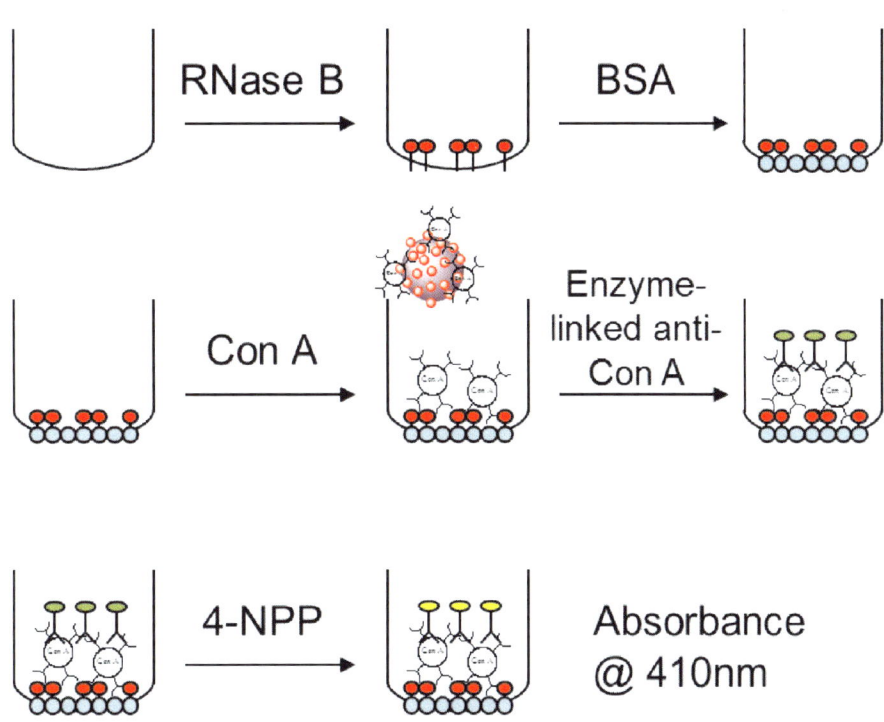

Figure 4.5. ELISA inhibition procedure. The plate was washed 3 times with PBS-T between each step.

ELISA inhibition experiments were performed on glycodendrimers of varying size and sugar loading (Figure 4.6 and Table 4.1). Data was fitted to a four-parameter dose-response curve to determine IC50 values. Monovalent methyl-mannose was added as a control to each ELISA plate to ensure results from separate plates were comparable. Compounds containing no mannose had no effect on the inhibition curve, suggesting that this interaction is specific and that the dendrimer architecture does not significantly add to non-specific binding. Compound **18g** also showed no inhibitory effect for the concentration ranges tested. For this compound, the few clusters are spread too far apart and are likely even buried in the interior of the large dendrimer.

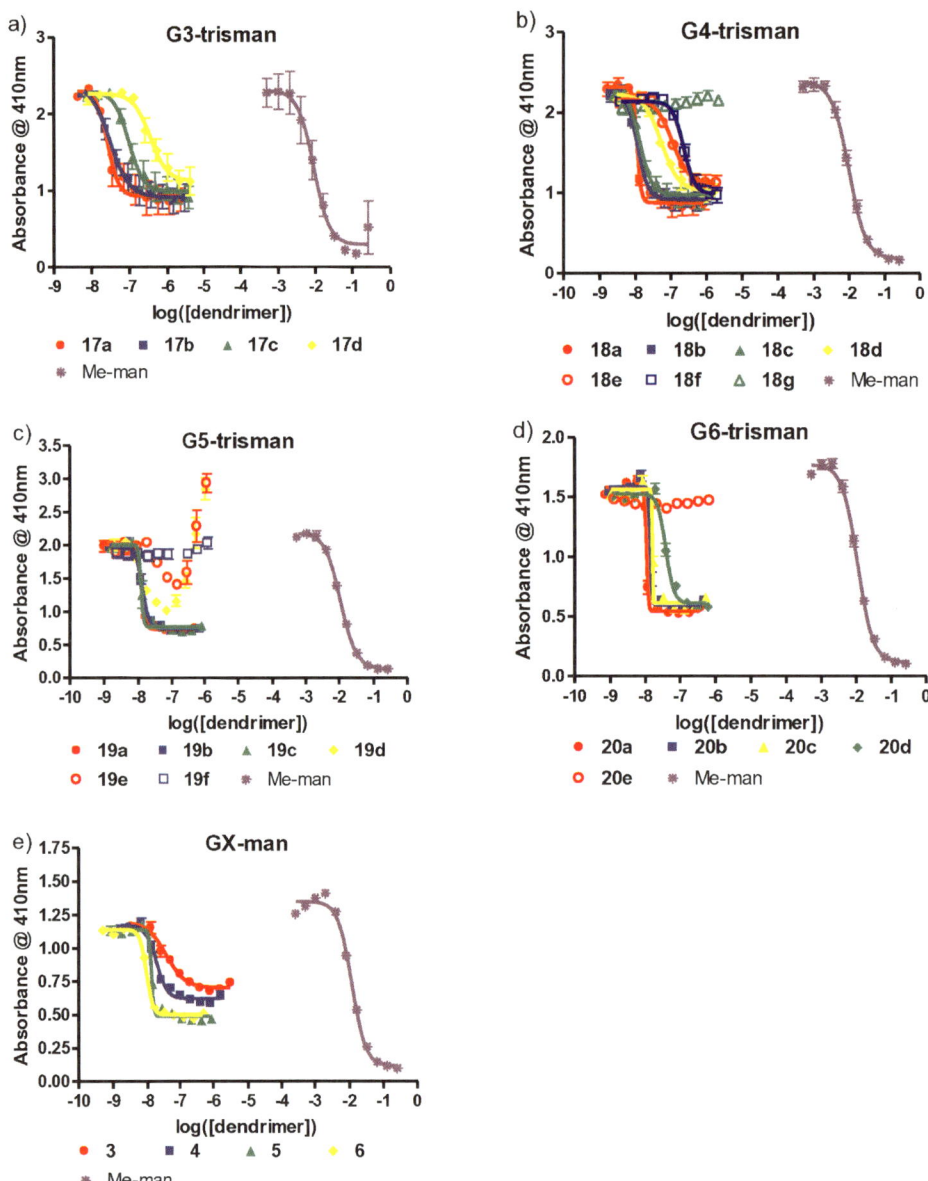

Figure 4.6. ELISA inhibition graphs for compounds **17-20** and **3-6**. The results are the averages of 3 replicate experiments.

Table 4.1. ELISA IC$_{50}$ values for tris-cluster dendrimers **17-20**. The values are the averages of 3 replicate experiments.

Dendrimer Generation	# Clusters	# Mannose	IC$_{50}$ (dendrimer)	IC$_{50}$ (mannose)
3 (**17a**)	13	39	27 nM	1100 nM
3 (**17b**)	8	24	30 nM	720 nM
3 (**17c**)	3	9	95 nM	860 nM
3 (**17d**)	1	3	350 nM	1100 nM
4 (**18a**)	39	117	11 nM	1300 nM
4 (**18b**)	26	78	11 nM	860 nM
4 (**18c**)	20	60	13 nM	780 nM
4 (**18d**)	9	27	54 nM	1500 nM
4 (**18e**)	5	15	110 nM	1700 nM
4 (**18f**)	3	9	230 nM	2100 nM
4 (**18g**)	2	6	N/A	N/A
5 (**19a**)	68	204	12 nM	2400 nM
5 (**19b**)	39	117	13 nM	1500 nM
5 (**19c**)	30	90	12 nM	1100 nM
5 (**19d**)	12	36	N/A	N/A
5 (**19e**)	8	24	N/A	N/A
5 (**19f**)	0	0	N/A	N/A
6 (**20a**)	66	198	11 nM	2200 nM
6 (**20b**)	31	93	14 nM	1300 nM
6 (**20c**)	22	66	15 nM	990 nM
6 (**20d**)	9	27	38 nM	1000 nM
6 (**20e**)	0	0	N/A	N/A

The tris-mannose cluster dendrimers were compared to non-clustered mannose functionalized dendrimers as a control (Figure 4.6 and Table 4.2). These dendrimers have been previously characterized and have their surfaces fully funcationalized.[52, 53, 105] Results show that presentation of the mannose endgroups in a clustered form does not provide a significant increase in activity when compared to a random, flexible presentation model. Comparison of the clustered and non-clustered mannose presentation scaffolds suggests that a denser clustering of the endgroups yields no significant increase in inhibition activity in a system as dynamic as a dendrimer (Figure 4.7). Both presentation models of mannose yield significant increases over monovalent methyl mannose. This observation is consistent with previous observations that suggests a small number of binding epitopes are able to display a maximum cluster effect.[157, 169] However, when the mannose endgroups are clustered as described above, the same number of mannose endgroups can be obtained while leaving sufficient space on the dendrimer for further functionalization.

Table 4.2. ELISA IC$_{50}$ values for mannose functionalized dendrimers **3-6** and monovalent methyl-mannose. The values are the averages of at least 3 replicate experiments.

Dendrimer Generation	# of sugars	IC$_{50}$ (dendrimer)	IC$_{50}$ (mannose)
3 (**3**)	29	37 nM	1100 nM
4 (**4**)	55	18 nM	990 nM
5 (**5**)	95	13 nM	1200 nM
6 (**6**)	178	9 nM	1600 nM
me-man	1	12 mM	12 mM

In two cases, compounds **19c** and **19d**, inhibition did not occur as expected. After an intial inhibition activity at lower concentrations, the signal increased rapidly, and even displayed activity above baseline. This suggests that with an increased amount of the glycodendrimer, Con A recruitment to the ELISA plate surface is promoted, rather than inhibited. The glycodendrimers are able bind to excess Con A, then carry them to the surface without being lost in subsequent washing steps. Presumably, compounds **19c** and **19d** are able to conform to the ideal size and shape needed to form a stable complex with a large amount of Con A while keeping the complex anchored to the Con A present on the surface of the ELISA plate, thereby linking a second layer of Con A to the surface of the ELISA plate. This reinforces the idea that size and shape complementarity plays a significant role in surface binding (cf. chapter 2).[176, 177, 178]

In general, higher surface loadings of mannose were more effective inhibitors on a per molecule basis, indicating an ability to more tightly bind and sequester Con A, preventing it from binding to the surface of the ELISA plate. Larger dendrimer generations are able to provide better inhibition on a per molecule basis, as they are able to carry more mannose endgroups (Figure 4.7a). However, on a per mannose basis, each dendrimer generation has an optimal activity at moderate surface loading (Figure 4.7b). This is due to steric interactions occurring at high loadings, from neighboring clusters preventing optimal access to individual mannose clusters. Furthermore, Con A is able to effectively block some clusters on the surface upon binding, effectively preventing access to them. Ideal inhibition occurs at moderate surface loadings, leaving enough endgroups available for potential imaging or therapeutic drug attachment. When comparing different generations of similar surface loadings, smaller generations inhibit significantly better when compared to their larger counterparts. Due to their smaller size and higher surface coverage, smaller dendrimers are presumably able to form more compact, stable complexes with Con A, better sequestering Con A from the surface of the ELISA plate.

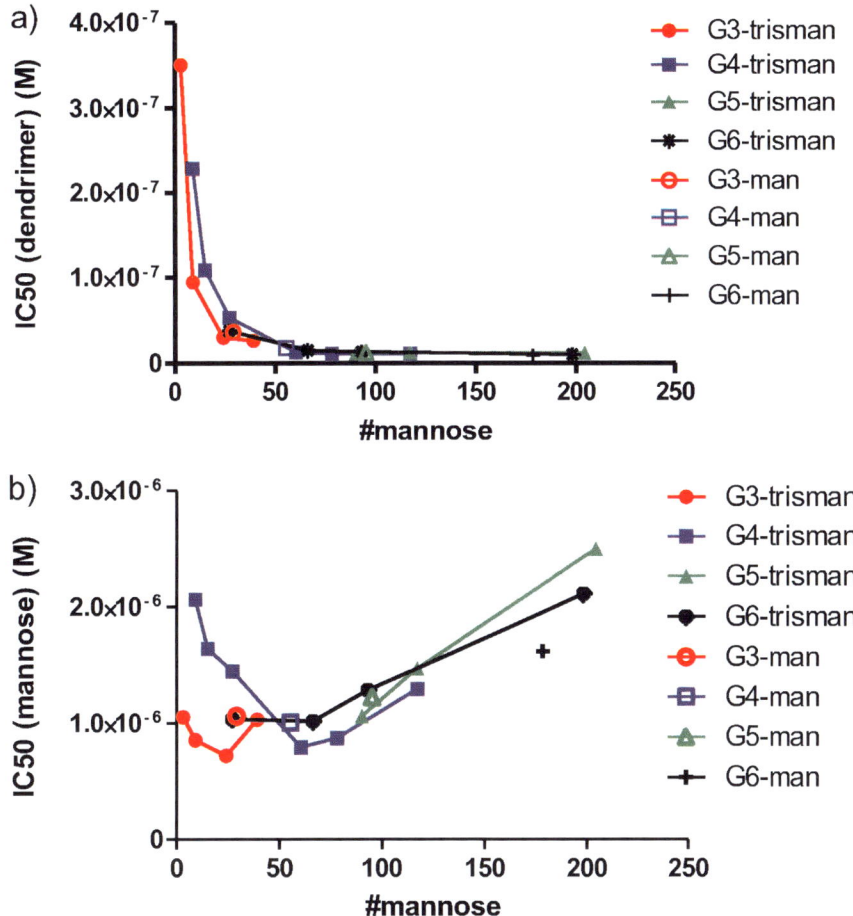

Figure 4.7. IC$_{50}$ values for **17-20** and **3-6** (a) represented on a per dendrimer basis and (b) represented on a per mannose basis.

Conclusions

Glycodendrimers bearing clustered mannose residues have been synthesized using click chemistry using a convergent synthesis approach. Use of click chemistry yielded glycosylated dendrimers **17-20** of varying sizes and surface patterning. ELISA inhibition experiments suggest that tris-clustered mannose residues show no significant difference in inhibition ability over single mannose endgroups when used on dynamic dendrimer scaffolds. Ideal inhibition occurs with smaller dendrimers when compared to larger dendrimers with comparable endgroups. Trismannose clustering allows for a redistribution of a scaffold's surface functionalities, potentially freeing up endgroups for other functionalities such as imaging or solubility.

Experimentals

General Methods. General reagents were purchased from Acros. Concanavalin A (Con A) was purchased from CalBioChem. Anti-Con A alkaline phosphatase conjugate was purchased from EY Laboratories (San Mateo, CA). Clustered mannose-functionalized dendrimers were prepared as described in reference 61. Non-clustered mannose-functionalized dendrimers were prepared as previously described (cf. reference 52, 53, 105).

MALDI-TOF MS. Matrix assisted laser desorption ionization (MALDI) mass spectra were acquired using a Bruker Biflex-III time-of-flight mass spectrometer. Spectra were obtained using a trans-3-indoleacrylic acid (IAA) matrix with a matrix-analyte ratio of 36000:1 to 1000:1. A 1 μL aliquot of the analyte (0.5-20 mg/mL) in DMF, DMA, or DMSO, was combined with 10 μL of IAA in DMF (20 mg/mL). A 1 μL aliquot was deposited on the laser target, drying in 1-4 hours. Bovine serum albumin (MW 66431 g/mol) and laser promoted oligomers up to the tetramer and the M/Z peak where Z = 2, trypsinogen (MW 23982 g/mol), cytochrome C (12361 g/mol), and bradykinin (1061 g/mol) were used as external standards. Positive ion mass spectra were acquired in linear mode, and the ions were generated by using a nitrogen laser (337 nm) pulsed at 5 Hz with a pulse width of 3 nanoseconds. Ions were accelerated at 19-20000 volts and amplified using a discrete dynode multiplier. Spectra (10 to 2000) were summed into a LeCroy LSA1000 high-speed signal digitizer. All data processing was performed using Bruker XMass/XTOF V 5.0.2. Molecular mass data and polydispersities of the broad peaks were calculated by using the Polymer Module included in the software package. The peaks were analyzed using the *continuous* mode. Delta values were set at minimum levels.

SDS-PAGE Assays with Glycodendrimers. SDS-PAGE assays were performed using a Glycoprotein Detection Kit (GLYCOPRO) obtained from Sigma. Ready Gels were obtained from Bio-Rad (Hercules, CA). SDS-PAGE analysis was performed using a Mini-PROTEAN 3 Cell (Bio-Rad) using pre-cast Ready Gels (Bio-Rad). 10% Ready Gels were used for higher molecular weight compounds (generation 5 and generation 6 dendrimers), and 15% Ready Gels were used for lower molecular weight compounds (generations 2, 3, and 4). Carbohydrates were detected using a modification of Periodic Acid-Schiff methods.[173, 174]

Compounds (100μg/mL – 1mg/mL) and standards (Precision Plus Protein Kaleidoscope Standards, Bio-Rad) were loaded onto gels together with loading buffer (Laemmli sample buffer, Bio-Rad). Horseradish peroxidase (HRP) was loaded as a positive control at a concentration of 1 mg/mL, reduced with β-mercaptoethanol and loaded with loading buffer. The samples were run at 200V for approximately 30 min.

After electrophoresis, the gels were fixed in 50% methanol for 30 minutes then washed twice with Millipore water for 10 minutes. The sugars present were oxidized by agitating the gel in a periodic acid solution for 30 minutes then washed twice with Millipore water for 10 minutes. Carbohydrates were stained by soaking the gel in a fuchsin-sulfite solution (Schiff's Reagent) until magenta bands were visible, or overnight. Unreacted carbohydrates were reduced by replacing the staining solution with a sodium metabisulfite reagent for 60 minutes. The gel was briefly washed with Millipore water then transferred to a 5% acetic acid storage solution or scanned for posterity.

Enzyme Linked Immunosorbent Assay (ELISA). Flat-bottomed 96-well microplates (Nunc MaxiSorp) were coated with 100 μL/well of 1 μg/mL RNase B in phosphate buffered saline (PBS) (pH 7.4, 10 mM phosphate, 150 mM NaCl), covered, and left at room temperature overnight. Plates then underwent a washing step using two washes with PBS-T (pH 7.4, 10 mM phosphate, 150 mM NaCl, 0.05% Tween-20) and one wash with PBS. Exposed surfaces were then blocked with 1% BSA in PBS-T, 100 μL/well, at 37 °C for 1 hour, then washed.

Dendrimer was dissolved into a phosphate buffered saline solution (PBS) (pH 7.4, 10 mM phosphate, 150 mM NaCl). The stock solutions for the glycodendrimer inhibitors were prepared at 100 μg/mL and the stock solution for methyl-mannose inhibitor was prepared at 100 mg/mL. 2-fold serial dilutions of the inhibitors were created in separate microcentrifuge tubes, to which an equal volume of 10 μg/mL Concanavalin A (Con A) in PBS-T were added. After 1 hour, 100 uL of each solution were transferred to the prepared ELISA plate. Positive controls containing no inhibitor and negative controls containing no Con A were also plated in this manner. The plates were incubated at 37 °C for 1 hour, then washed as above.

100 μL of Anti-Con A alkaline phosphatase conjugate (in PBS-T, 1% BSA) was added to each well and incubated at 37 °C for 1 hour, then washed. 100 uL/well of p-nitrophenyl phosphate was added, and left at room temperature until as strong yellow color presented itself (~30 minutes). Further conversion was stopped with 100 uL 1M NaOH as a stopping solution in each well. Absorbances were taken at 410 nm using a reference wavelength of 540 nm. IC50 values were obtained by fitting data to a 4-parameter dose-response equation in GraphPad Prism 4 (La Jolla, CA).

CHAPTER 5

CHARACTERIZATION OF PROTEIN AGGREGATION VIA INTRINSIC
FLUORESCENCE LIFETIME

Introduction

The evaluation of protein-carbohydrate interactions has relied heavily on qualitative assays such as hemagglutination and precipitation assays.[114, 179] More quantitative approaches such as isothermal titration microcalorimetry, surface plasmon resonance and fluorescence anisotropy have provided more informative results. However, these methods are often unsuitable for the study of large, multivalent frameworks because precipitation is prevalent under the conditions required for the experiments.[116, 117, 180]

Aggregation plays an integral role in many cellular pathways, one of the most important being mediation of the infection and proliferation potential of tumors and pathogens.[181] Protein aggregation has also been implicated in pathological conditions such as Alzheimer's and other amyloid-related diseases.[182] Because of the importance of multivalently displayed carbohydrates on cell surfaces, sugar-induced aggregation has drawn considerable attention.[12, 13, 41, 42, 183, 184, 185] Sensor strategies based on controlled aggregation have been reported for the detection of toxins and other biologically relevant compounds.[186, 187, 188, 189, 190] Multivalent interactions often involve multiple weak monovalent binding events. An in-depth understanding of aggregation in complex systems requires studies that go beyond measuring the monovalent association constants.

Particularly valuable would be methods capable of characterizing aggregation events in real time.

Measurements based on a protein's intrinsic fluorescence lifetime properties offer a method of observing binding that is less sensitive to light source intensity or scattering effects, which are often an issue in systems where extensive aggregation and precipitation occur.[191] Furthermore, using the intrinsic tryptophan fluorescence of proteins eliminates the need for labeling, offering a rapid measurement of events in biological conditions. Changes in the fluorescence lifetime can also indicate changes in the aggregation state, which influences overall immunogenicity.[192]

In this study, the aggregation of the mannose-specific lectin Concanavalin A by glycodendrimers is examined. Sugar-functionalized dendrimers (Figure 5.1) provide a controlled, synthetic scaffold that is ideal for studying and mediating multivalent interactions. Several studies of glycodendrimers binding to Concanavalin A (Con A) were previously reported, revealing binding and inhibition efficacies dependent on size and functionalization.[52, 53, 105] Here, a novel approach using fluorescence lifetime measurements of mannose-functionalized dendrimers is described, providing an assay for the evaluation of glycodendrimer/protein complexes, even in solutions where precipitation is occurring.

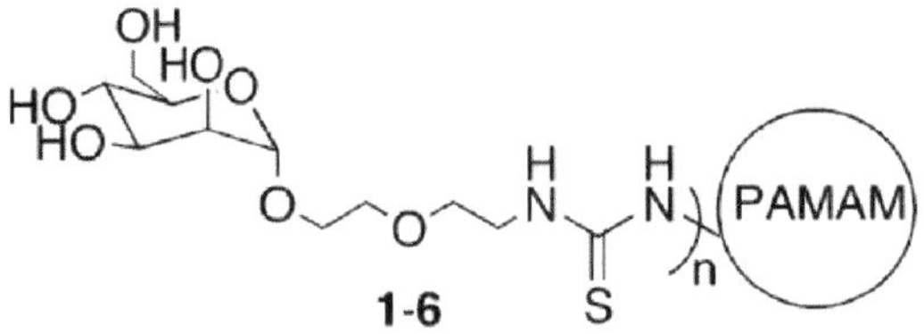

1 G(1); n = 8
2 G(2); n = 16
3 G(3); n = 29
4 G(4); n = 55
5 G(5); n = 85
6 G(6); n = 172

Figure 5.1. Mannose-functionalized PAMAM dendrimers.

Fluorescence Lifetime

Fluorescence occurs when an electron of a molecule is excited above its ground-state orbital (S_0) into an excited singlet state (Figure 5.2).[191] Through internal conversions and vibrational relaxations, the electron returns to the lowest vibrational energy level of the excited state before emitting a photon. Return to the ground state happens with the emission of a photon at a typical decay time of 1-10 nanoseconds. Molecules in the S_1 state can also undergo a spin conversion to change their orientation to an energetically lower T_1 triplet state through intersystem crossing. Emission of light from a triplet state, in which the two electrons have the same spin orientation as the electron in the ground state, is termed phosphorescence. Transition from the triplet state to the ground state in the same molecule is forbidden by the Pauli exclusion principle. Because of this, the rate constants for emission of a photon are much smaller than those for fluorescence, leading to longer lifetimes in milliseconds to seconds.

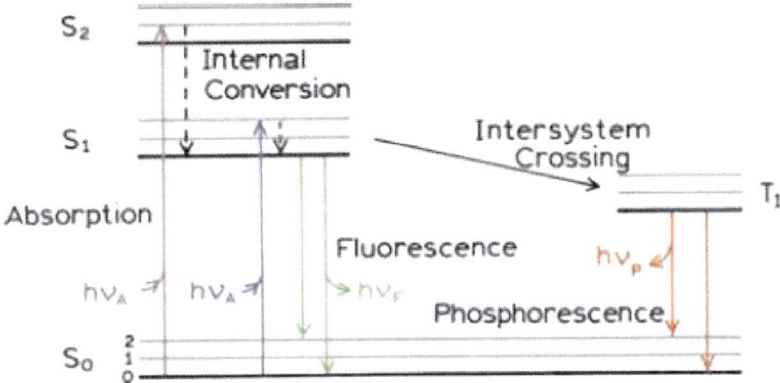

Figure 5.2. Jablonski diagram describing fluorescence. (Adapted from reference 191)

Because of the short timescale, measurement of fluorescence lifetimes requires sophisticated optics and electronics. However, time-resolved fluorescence offers a great deal of information when compared to steady-state, intensity-based measurements. Intensity measurements are an average of fluorescence lifetimes, and the averaging process disregards the information contained within the shape of the fluorescence decay curve.[191] Macromolecules exist in several single conformations, and the fluorescence lifetime of the molecule may depend on the conformation in which it is present. The population of the fluorescence lifetimes could reveal the presence of multiple conformational states, whereas an intensity measurement will only reveal the average of the fluorescence intensities present.

Extrinsic fluorophores are fluorophores which are added to a sample in order to give the sample fluorescent properties. Intrinsic fluorophores are fluorophores which naturally occur in a molecule or system. In proteins, the indole group of tryptophan lends itself well to intrinsic fluorescence studies. Indole absorbs light near 280 nm, and emits near 340 nm. It is highly sensitive to solvent effects and is affected by immediate environmental effects, such as when the protein is folded in its native state or unfolded.[191]

Tryptophan is a dominant intrinsic fluorophore, present in proteins at about 1%.[191] The relatively low abundance of tryptophan lends itself well for intrinsic fluorescence studies, facilitating the interpretation of spectral data without having too many signals to deconvolute. Because tryptophan residues are highly sensitive to their immediate environment, this chapter uses the intrinsic fluorescence from the indole ring to examine Concanavalin A binding induced by a multivalent glycodendrimer system.

Con A Complexation Measured by Intrinsic Fluorescence

To assess the binding behavior of Con A with glycodendrimers, the changes in the fluorescence lifetime waveform were monitored with subsequent additions of glycodendrimer to a solution of Con A. Briefly, known amounts of glycodendrimer solution were added to 2000 μL of 100 μg/mL Con A in a well-stirred cuvette at 25 °C. A baseline of Con A fluorescence was established for 30 s before a glycodendrimer aliquot was added. Fluorescence decay waveforms were measured once per second for the next 130 to 220 s. Using the initial waveform of free Con A and the final waveform of complexed Con A as basis waveforms, the waveforms of each addition were fit as a linear combination of free and complexed Con A basis waveforms. This expresses the complex formation induced by the glycodendrimer as the amount of Con A complexed with time (Figures 5.3 – 5.6).

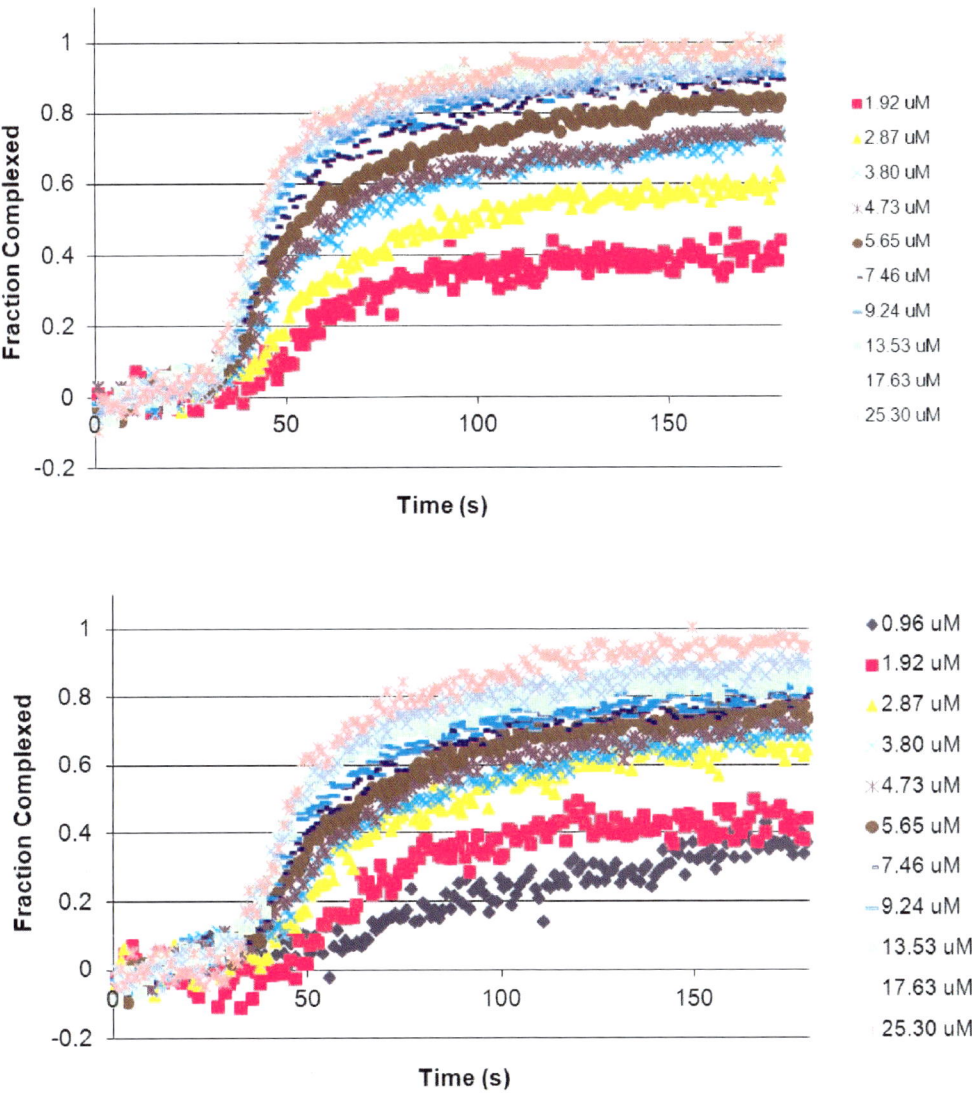

Figure 5.3. Fluorescence assay data for additions of G(2)-man **2** into 100ug/mL Con A.

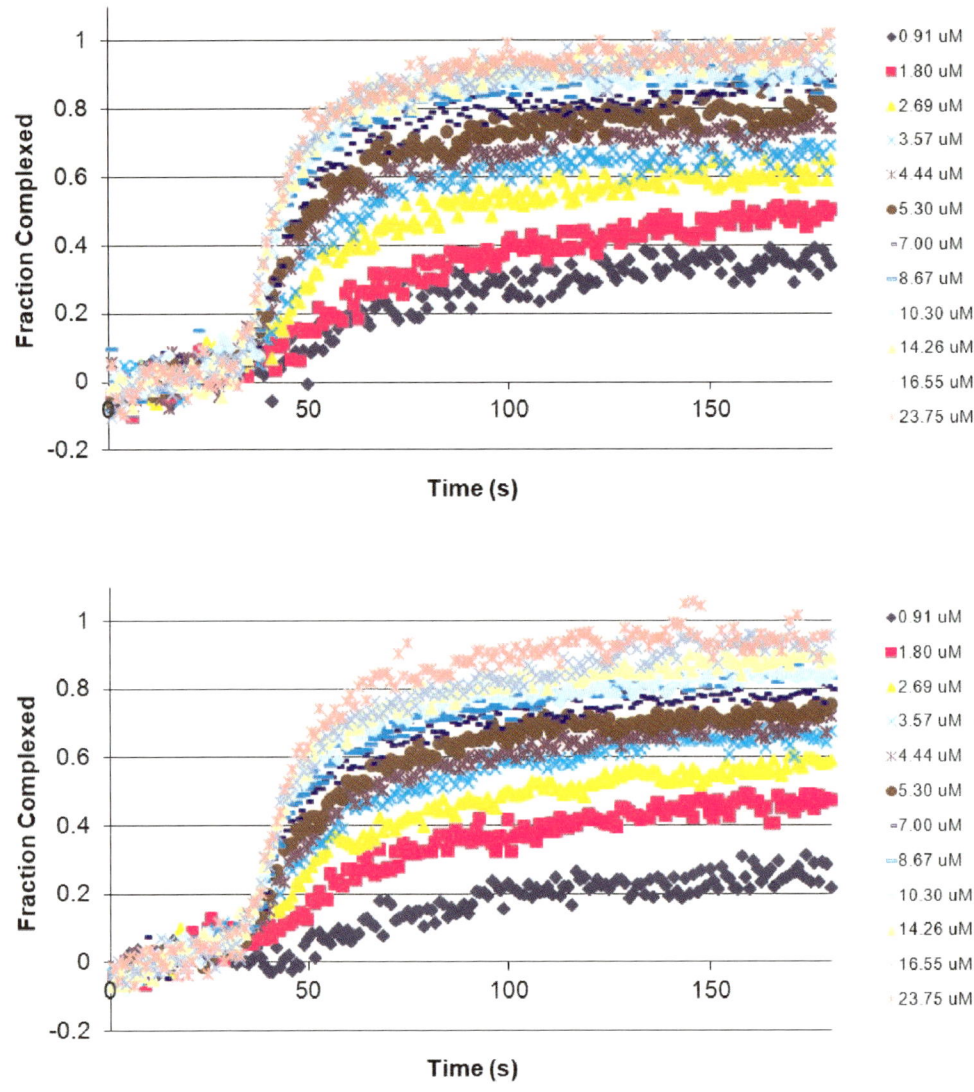

Figure 5.4. Fluorescence assay data for additions of G(3)-man **3** into 100ug/mL Con A.

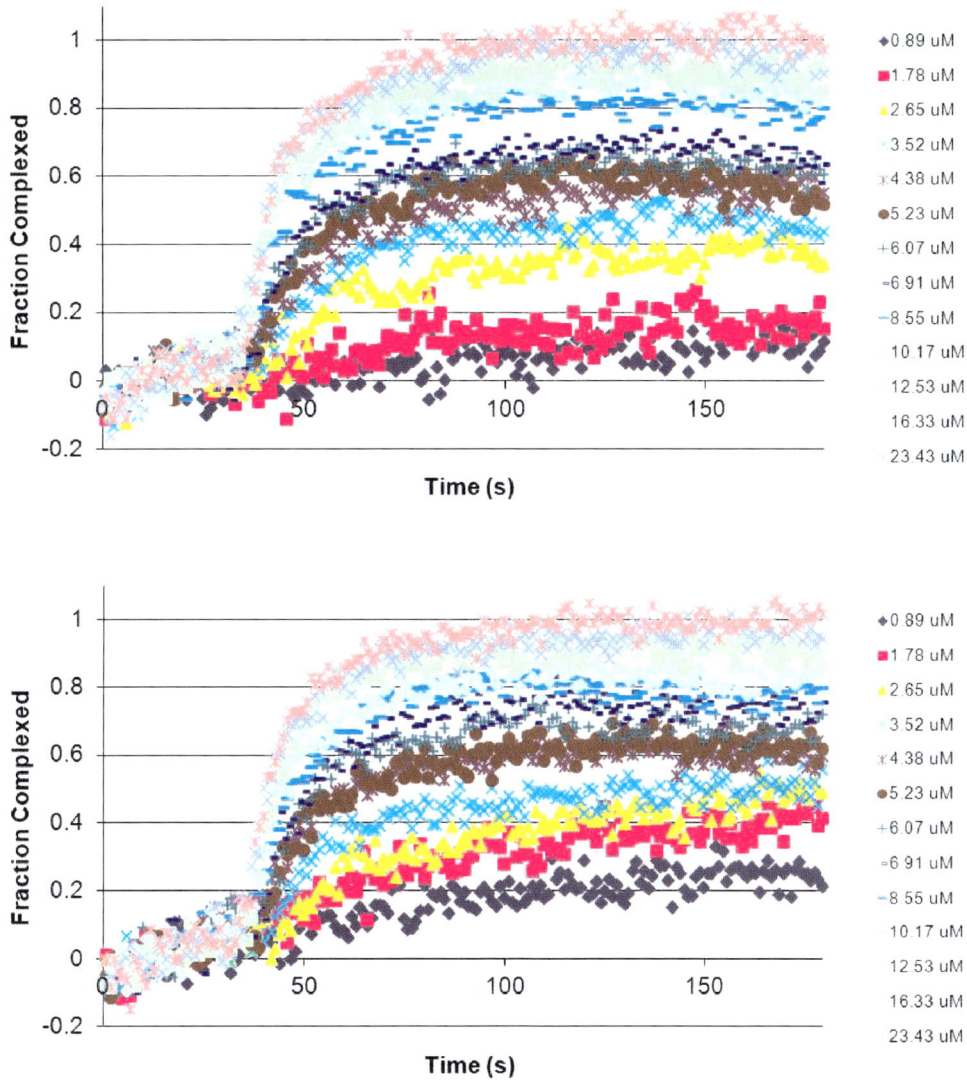

Figure 5.5. Fluorescence assay data for additions of G(4)-man **4** into 100ug/mL Con A.

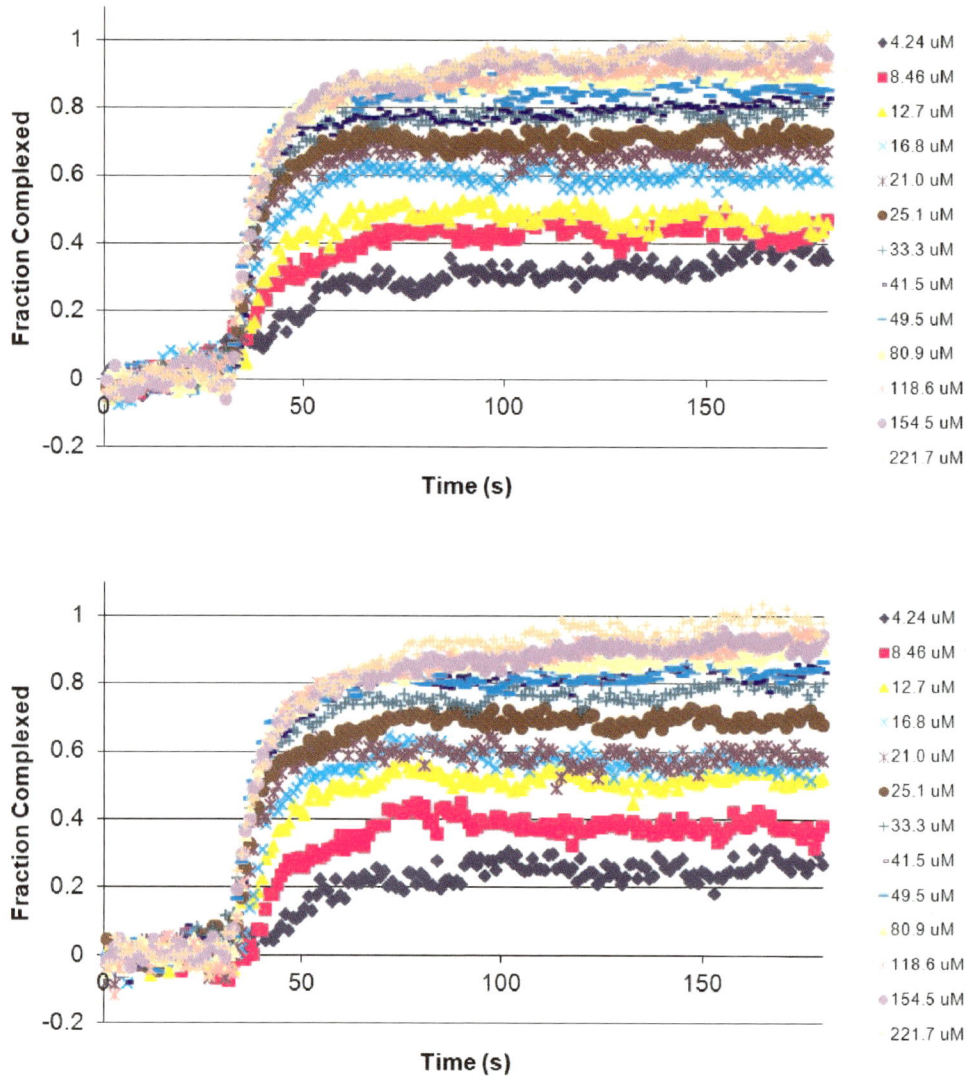

Figure 5.6. Fluorescence assay data for additions of G(6)-man **6** into 100ug/mL Con A.

Control experiments showed that the primary sugar-Con A binding events are not the source of the observed fluorescence changes. Several millimolar concentrations of R-*O*-methyl mannoside (Me-man) (Figures 5.7 and 5.9) had minimal effect on the Con A fluorescence and did not result in precipitate formation. To ensure that the dendrimer framework itself was not the source of the Con A quenching, a galactose-functionalized dendrimer that does not bind to Con A was added under the same conditions as the mannose-functionalized dendrimers **2-4** and **6**. Again, no precipitation was observed visually, and nonspecific binding was minimal relative to baseline drift and noise. This suggests that the waveform differences are arising from the mannose-functionalized dendrimers inducing a conformation change of the system and are not due to the sugar binding or dendrimer framework alone (Figures 5.8 and 5.9).

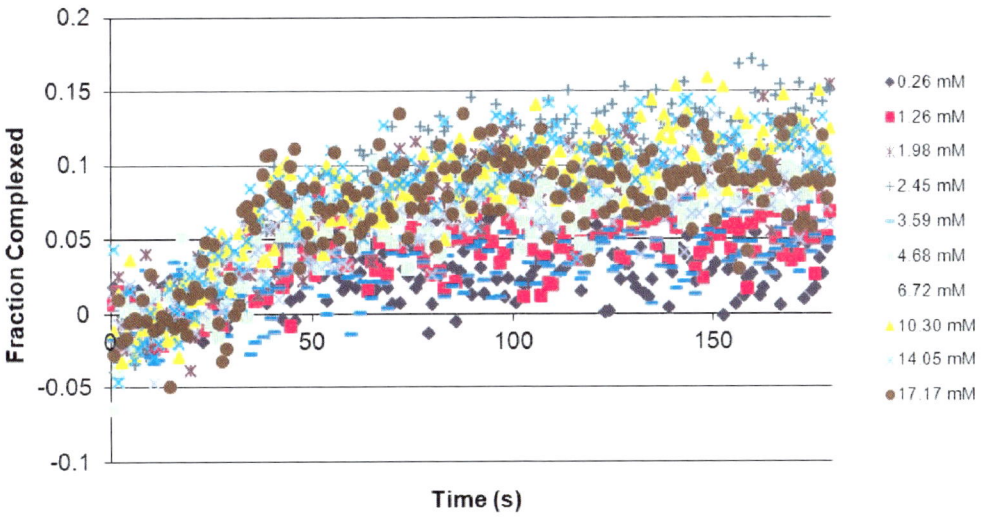

Figure 5.7. Fluorescence assay data for additions of Me-man into 100ug/mL Con A.

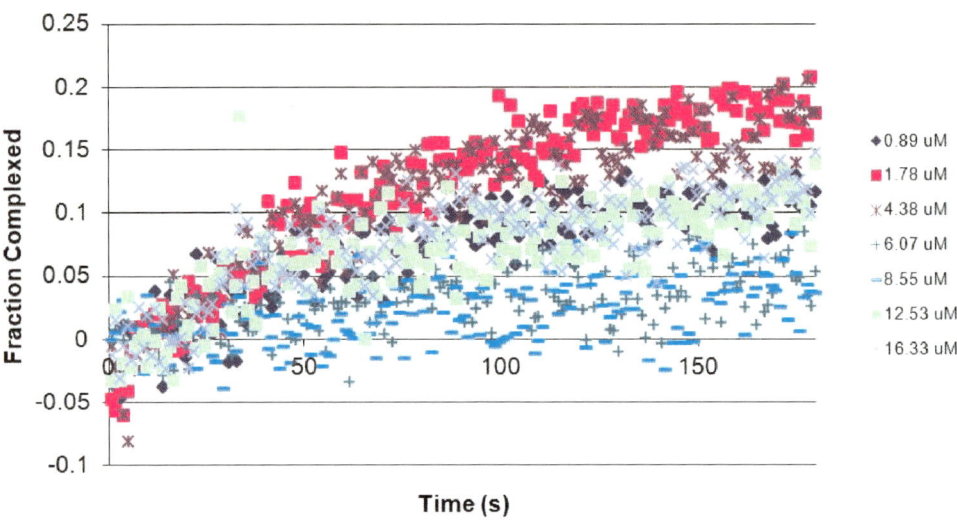

Figure 5.8. Fluorescence assay data for additions of G(4)-gal into 100ug/mL Con A.

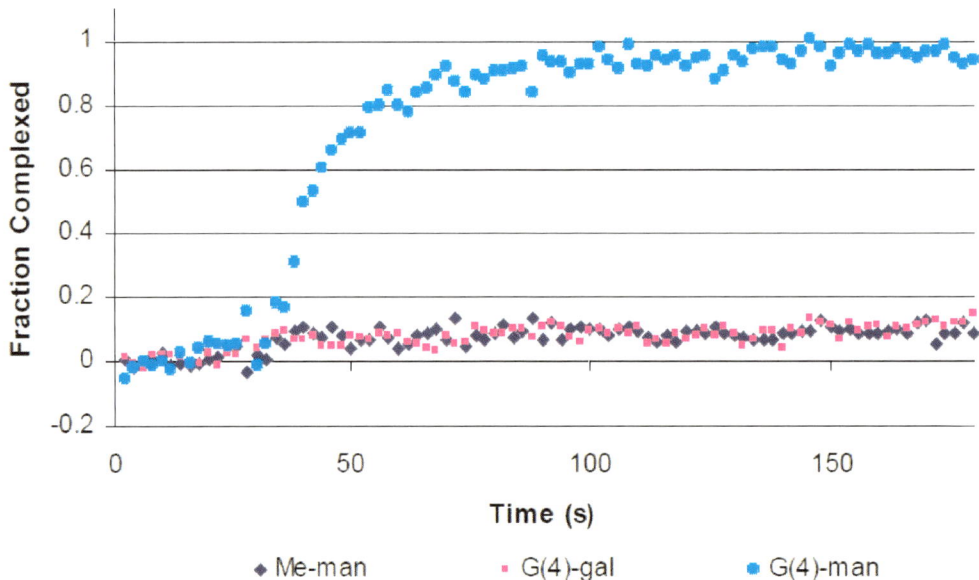

Figure 5.9. Complex formation upon addition of 17.2 mM Me-man, 16.3 μM G(4)-gal and 16.3 μM G(4)-man **4**.

The fluorescence changes are most likely associated with a Con A-Con A protein-protein interaction orchestrated by binding to the glycodendrimer framework. Figure 5.10 illustrates our hypothesis that a reversible protein-protein interaction between proximal Con A lectins causes the quenching. Con A possesses 4 tryptophan residues per monomeric unit, of which Trp88 is on the outer surface of the protein (Figure 5.11). With an increased local concentration of Con A, Trp88 could easily be affected by protein crowding. The primary role of the dendrimers is to hold the Con A lectins in close proximity, thereby increasing their effective concentration. A protein-protein interaction, presumably a conformational change affecting the environment of one or more Con A tryptophans, can then occur.

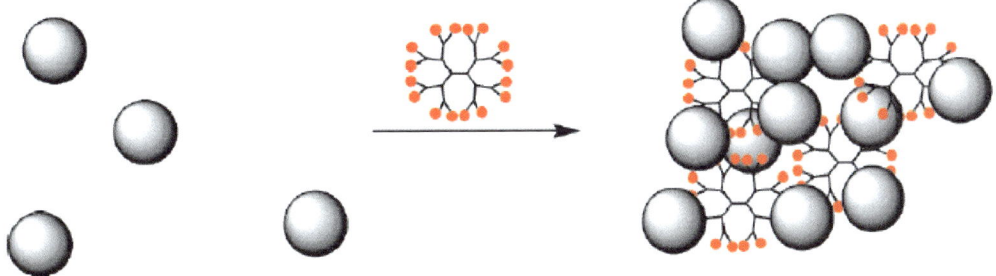

Figure 5.10. Glycodendrimer-mediated lectin aggregation: (left) uncomplexed Con A; (right) cross-linked state.

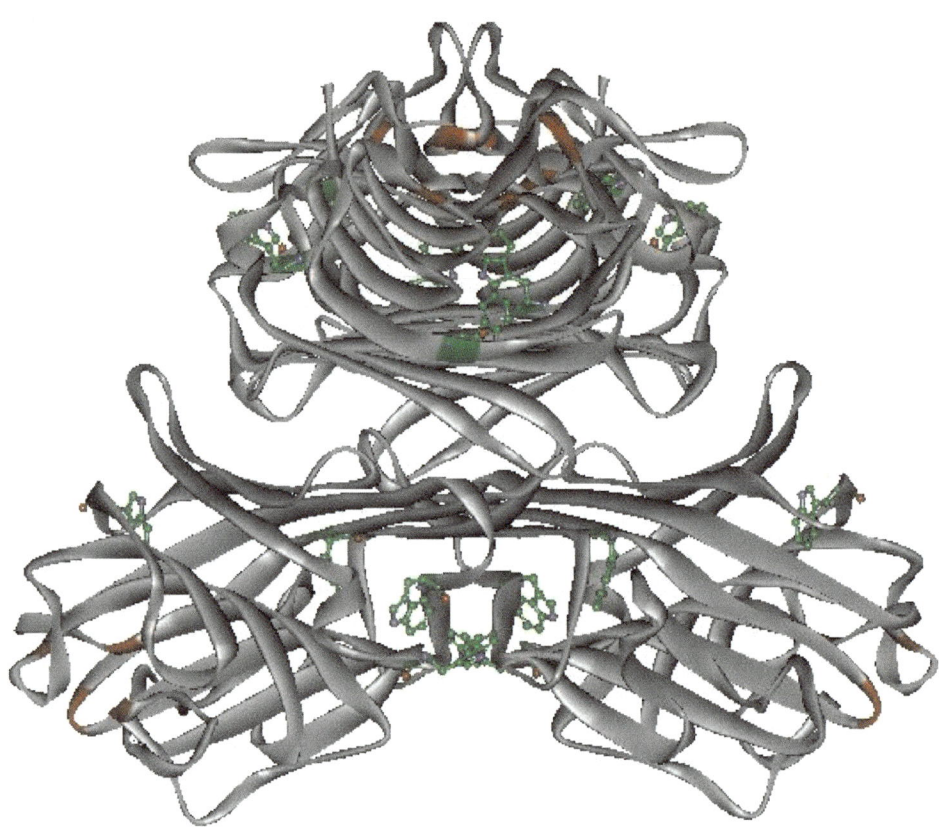

5.11. Structure of Con A with active binding sites highlighted in red and tryptophan residues highlighted in green.[32]

All generations of glycodendrimer showed saturation behavior, i.e., the fluorescence changes reached a plateau at sufficiently high dendrimer concentration. Furthermore, the fluorescence decay curves at saturation were identical within the experimental uncertainty for all dendrimer generations. Thus, all of the fluorescence decay curves for the entire range of dendrimer concentrations and all generations could be fit to a linear combination of just two waveforms, corresponding to "free" and "complexed" Con A. The environment of the complexed Con A is apparently not affected by the generation of dendrimer (Figure 5.12).

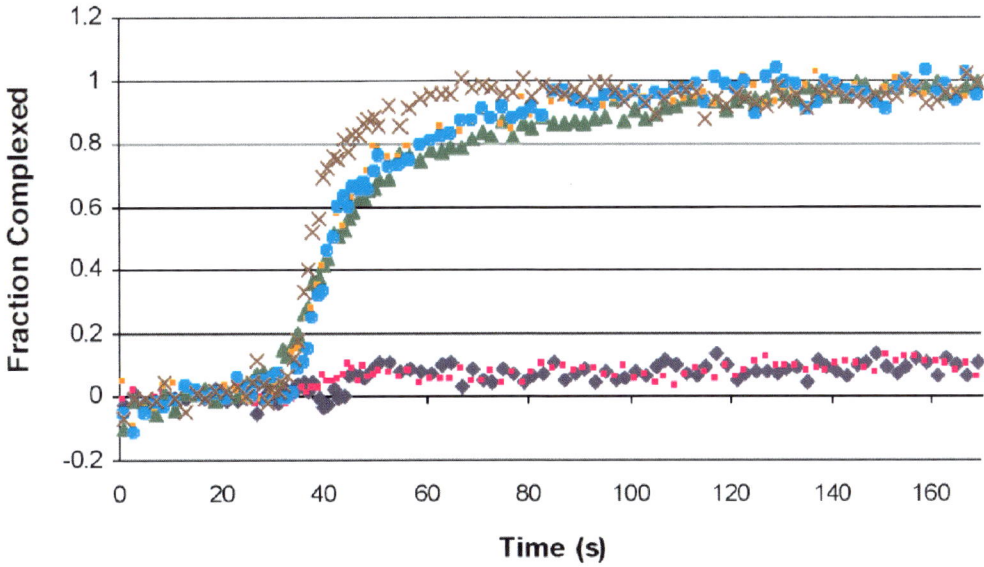

Figure 5.12. Complex formation of Con A with time as a result of dendrimer addition. 1-O-methyl mannoside (♦, 17.2 mM), galactose-functionalized G(4)-PAMAM (■, 16.3 μM), **2** (▲), **3** (■), **4** (●), **6** (✕) (all 21 to 25 μM). Data for mannose-functionalized dendrimers were normalized to the same endpoint to emphasize rate differences between glycodendrimer generations.

Fitting the appearance of the complexed state to a 1:1 binding model affords an apparent kinetic rate constant k_{obs} indicative of the ability of the dendrimers to sequester Con A lectins and bring them into proximity with one another (Figures 5.13 and 5.14, Table 5.1). It is not implied that a 1:1 binding model is in effect; rather, the comparison of k_{obs} is a convenient way to observe how differences in dendrimer generation affect glycodendrimer mediated protein aggregation. Glycodendrimers **2** and **3** offer similar trends in k_{obs}, while **6** yields a 3- to 4-fold increased apparent rate constant on a per sugar basis. As has been the case with previous assays,[52, 53, 105] results with **4** fall between the those for large and small dendrimers.

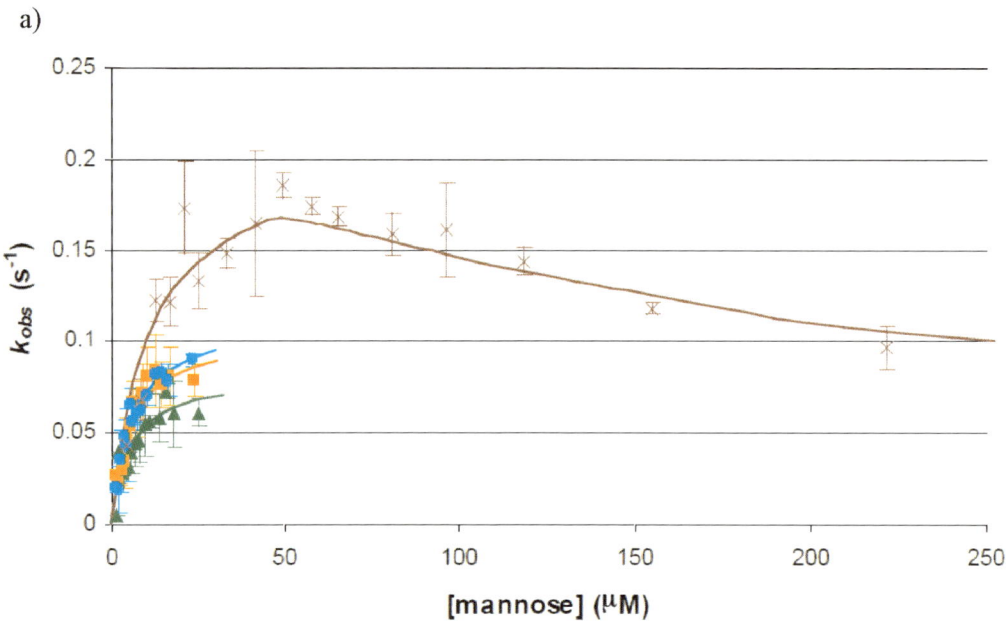

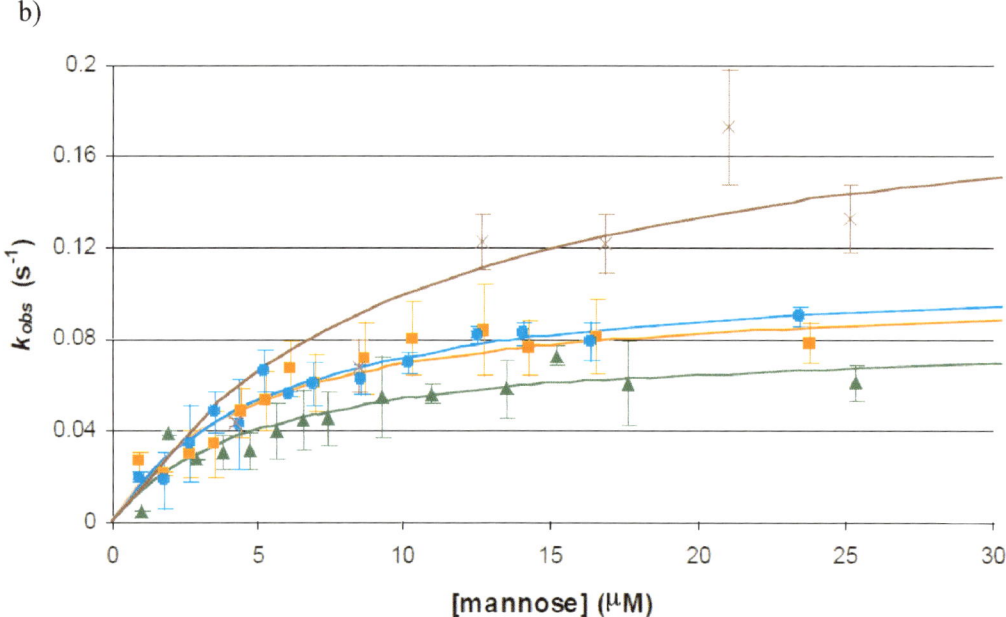

Figure 5.13. (a) Kinetic data for compounds **2-4** and **6**. k_{obs} presented on a mannose concentration basis. (b) Expanded view. **2** (▲), **3** (■), **4** (●), **6** (✕)

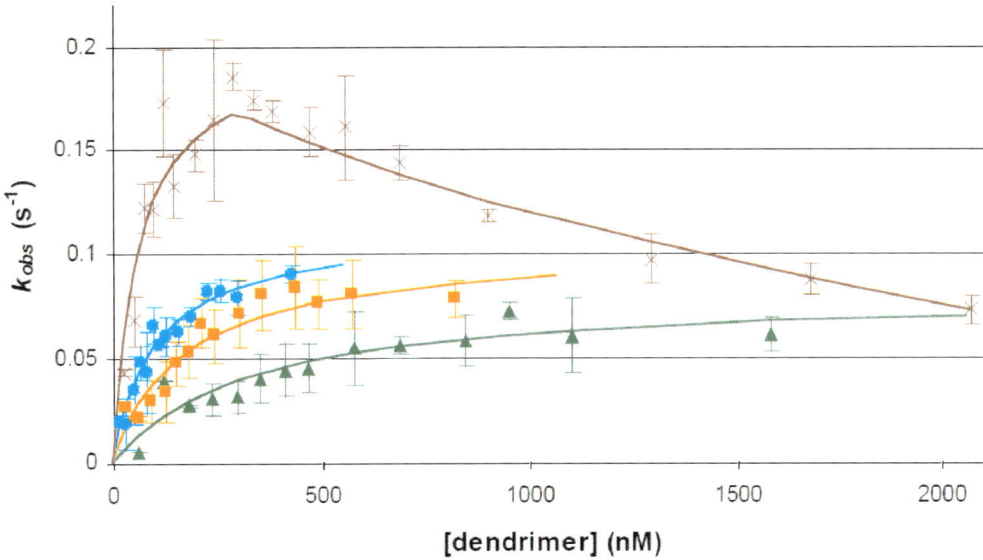

Figure 5.14. Kinetic data viewed in terms of dendrimer concentration. **2** (▲), **3** (■), **4** (●), **6** (✕)

Table 5.1. Calculated k_{obs} for compounds **2-4** and **6**.

G(2)		k_{obs} (sec^{-1})		G(3)		k_{obs} (sec^{-1})	
[dendrimer] (nM)	[sugar] (μM)	Trial 1	Trial 2	[dendrimer] (nM)	[sugar] (μM)	Trial 1	Trial 2
60	0.96	0.0097	0.0022	31	0.91	0.0136	0.0196
120	1.92	0.0262	0.0200	62	1.80	0.0154	0.0200
179	2.87	0.0203	0.0201	93	2.69	0.0288	0.0241
238	3.80	0.0269	0.0200	123	3.57	0.0365	0.0250
296	4.73	0.0309	0.0233	153	4.44	0.0379	0.0288
353	5.65	0.0295	0.0249	183	5.30	0.0383	0.0337
410	6.56	0.0335	0.0262	212	6.16	0.0460	0.0382
466	7.46	0.0322	0.0301	241	7.00	0.0490	0.0367
577	9.24	0.0389	0.0320	299	8.67	0.0503	0.0343
686	11.0	0.0377	0.0336	355	10.3	0.0528	0.0461
846	13.5	0.0443	0.0369	438	12.7	0.0570	0.0419
950	15.2	0.0418	0.0476	492	14.3	0.0496	0.0444
1102	17.6	0.0447	0.0361	571	16.5	0.0543	0.0409
1581	25.3	0.0435	0.0366	819	23.7	0.0540	0.0442

G(4)		k_{obs} (sec^{-1})		G(6)		k_{obs} (sec^{-1})	
[dendrimer] (nM)	[sugar] (μM)	Trial 1	Trial 2	[dendrimer] (nM)	[sugar] (μM)	Trial 1	Trial 2
16	0.89	0.0131	0.0111	25	4.24	0.0440	0.0377
32	1.78	0.0151	0.0167	49	8.46	0.0598	0.0633
48	2.65	0.0296	0.0217	74	12.65	0.0849	0.0808
64	3.52	0.0345	0.0291	98	16.83	0.0956	0.0877
80	4.38	0.0355	0.0380	122	20.98	0.1043	0.1258
95	5.23	0.0434	0.0399	146	25.12	0.1071	0.0870
110	6.07	0.0443	0.0466	194	33.33	0.0914	0.0952
126	6.91	0.0411	0.0451	241	41.45	0.1031	0.0842
156	8.55	0.0439	0.0521	288	49.50	0.1052	0.0973
185	10.2	0.0484	0.0509	334	57.5	0.0898	0.0897
228	12.5	0.0500	0.0561	380	65.4	0.0976	0.0921
256	14.1	0.0485	0.0536	471	80.9	0.0920	0.0854
297	16.3	0.0503	0.0536	559	96.2	0.0915	0.0843
426	23.4	0.0552	0.0558	689	119	0.0772	0.0774
				898	155	0.0751	0.0684
				1289	222	0.0636	0.0558

These results suggest that **2** and **3** behave similarly when binding Con A, which is as expected since these glycodendrimers are both too small to multivalently bind to Con A. The Con A binding sites are approximately 6.5 nm apart, and **6** has a diameter of 13 nm. Thus, about 16% of the circumference of **6** (a reasonable value for divalent binding) is required to span the Con A binding sites. The 6 nm diameter of **3** dictates that one third of the circumference must be spanned in a divalent binding scenario, and this is sterically unrealistic. Due to its ability to bind Con A multivalently, **6** has previously been shown to exhibit significant increases in activity with Con A.[52, 53, 105] The apparent rate constant reflects the ability of the mannose-functionalized dendrimers to hold lectins in close proximity to one another, with **6** most effective but with all generations forming significant and similar aggregates.

The rate of complex formation is sensitive to the dendrimer generation and concentration. The kinetics traces (Figures 5.13 and 5.14) often appeared biphasic, with the much faster first step generally accounting for >90% of the total change. Except for very low dendrimer concentrations, a first-order kinetics model fit the data very well for the first 70 s after dendrimer addition and yielded an apparent rate constant k_{obs}.

Using a global fitting model, values for k_{on} and k_{off} could be obtained by taking the concentration dependence of k_{obs} into account. Assuming the fluorescence change arises from a single type of interaction, namely protein-protein proximity, kinetic data can be determined from the observed rate changes with different glycodendrimer concentrations (Table 5.2).

Table 5.2. Kinetic data determined from global fits of k_{obs}.

	k_{on} (M^{-1} s^{-1}) (dendrimer basis)	k_{on} (M^{-1} s^{-1}) (sugar basis)	k_{off} (s^{-1})
2	$21 * 10^3$	$1.3 * 10^3$	$28 * 10^{-3}$
3	$68 * 10^3$	$2.3 * 10^3$	$30 * 10^{-3}$
4	$140 * 10^3$	$2.5 * 10^3$	$39 * 10^{-3}$
6	N/A	N/A	N/A

These rate constants confirm that glycodendrimers **2** and **3** bind Con A similarly, while **6** binds Con A in a more complex binding motif. Of the dendrimers tested, **4** is able to bind Con A multivalently, while **2** and **3** are much too small to span multiple binding sites found on Con A. **6** does not fit a 1:1 binding model, which suggests that the multivalent interaction is adding an additional degree of complexity when compared to the other glycodendrimers. Interestingly, the increased rate observed from adding glycodendrimer does not arise from a large change in the dissociation rate, even though the local concentration of ligands available is greatly increased as the generation is increased. Instead, the observed rate constant is primarily the result of a significantly increased association rate.

Conclusions

Although the mechanism is undoubtedly much more complicated than a simple first-order process, k_{obs} is a convenient way to depict how differences in dendrimer generation affect glycodendrimer-mediated protein aggregation. Comparing the results across dendrimer generations revealed an increase in the apparent rate constant with increasing dendrimer generation expressed on a per-mannose basis. At low concentrations of added dendrimer, there is a fixed pool of Con A lectins competing for a small number of glycodendrimers. The rate-limiting step in achieving a complex that changes the fluorescence decay waveform may be two Con A lectins becoming cross-linked into close proximity by a small number of glycodendrimers. Thus, the overall apparent rate constant increases as the concentration of glycodendrimer increases. As expected, the rate constant approaches a saturation value at sufficiently high dendrimer concentration, at which the rate-limiting step becomes the Con A-Con A mutual conformational change.

The multivalent binding in **6** boosts the kinetics over what is observed for the lower-generation dendrimers because it also serves to increase the Con A residence time on a sugar binding site. However, a slowing of the kinetics for very high dendrimer concentration in G6 can be seen. The slower kinetics at higher G6 concentrations can be understood as a consequence of some fraction of the Con A lectins initially being too far apart to undergo the protein-protein interaction (each Con A has more sugar binding sites to choose from) and the multivalent binding impeding their ability to move to new sites.

Because the same final degree of complex formation is achieved, it is suggested that there is a slower process by which the Con A's migrate into adjacent positions on the dendrimer framework, which is the most stable configuration. Brewer and co-workers[193] have proposed a model in which lectin molecules bind and jump from carbohydrate epitope to epitope. It is proposed that an attractive interaction between Con A molecules on adjacent glycodendrimer binding sites exists that causes their residence times to increase substantially. Notably, once the entropic penalty of Con A-sugar binding has been paid, only a small enthalpic stabilizing force is necessary for the protein-protein interaction. It is also interesting to speculate on a possible connection between our observations and the selectivity switching reported by the Whitesides group.[194]

In summary, fast and precise fluorescence lifetime experiments were performed using unlabeled lectin to characterize glycodendrimer-mediated protein aggregation.

Lifetime measurements were used in these experiments to explain self-quenching phenomena induced by aggregation states, but this method is not limited to such and is viable for numerous binding studies. For example, intrinsic fluorescence can be used for the study of protein-protein interactions, protein-small-molecule interactions, vesicle and micelle formation, oligomerization events, and protein folding. Labeled compounds can be studied as well using the methods described here.

Instrumentation

Intrinsic fluorescence is attractive because of its label-free aspects, but light scattering and inner filter effects associated with the extensive precipitation that often accompanies aggregation hamper steady-state approaches. Fluorescence lifetime approaches are often stated to be immune to precipitation problems because the desired information can be extracted from the shape of the decay curve rather than the intensity.[191, 192] However, conventional lifetime technology is too slow for studying reactions with half-lives as short as just a few seconds. The data reported here were collected with a prototype instrument that increases by a factor of ~100 the rate at which fluorescence lifetime data can be collected.

Conventional fluorescence lifetime acquisition detects only a single photon out of 100 pulses due to the electronics of the instruments.[191] To achieve a noise level of 1%, 10,000 photons must be counted from 1,000,000 pulses within a short timeframe. This noise level is the result of the uncertainty associated with photon quantum statistics. In conventional fluorescence lifetime experiments, this data collection method is limited by the photodetector and digitizer, which limits the time resolution to the nanosecond range.

Fluorescence Innovations, Inc. has developed a proprietary digitizer capable of recording the entire decay curve for each pulse, instead of being limited to counting single photons. This enables the use of lower repetition frequencies and more powerful lasers, resulting in more precise data acquisition at speeds hundreds of times faster than conventional fluorescence lifetime instruments. The increased speed and higher precision data allows for fluorescence lifetime measurements in experiments impossible through previous methods.

Fluorescence data was collected on a Varian Eclipse spectrometer that Fluorescence Innovations, Inc. modified for fluorescence lifetime measurements. The modifications involved: (1) addition of a tunable UV source (components 1-3, Figure 5.15) as an alternative to the usual flashlamp excitation source; (2) installation of a R7400 photomultiplier tube (PMT)(component 4, Figure 5.15) in addition to the standard R928 PMT (component 5, Figure 5.15), and (3) introduction of a transient digitizer to capture time domain information on the nanosecond time scale. It should be noted that introduction of the laser source and the alternative PMT was accomplished without impacting the standard functionality of the Eclipse. It was also equipped with a TLC 50 cuvette holder (Quantum Northwest, WA) for temperature control and magnetic stirring capabilities. A schematic drawing of the modified Eclipse is shown in Figure 5.15. The components outlined in blue were not utilized in making fluorescence lifetime measurements but are retained for standard Eclipse functionality.

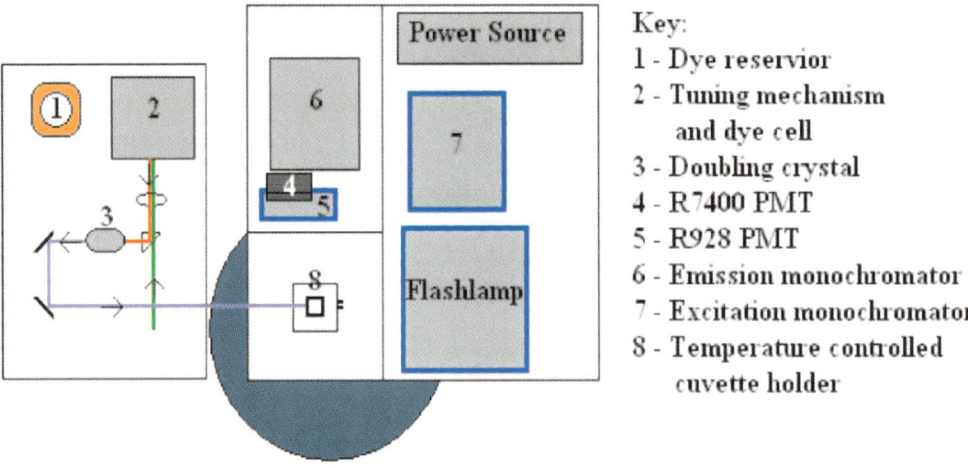

Figure 5.15. Diagram of Varian Eclipse Spectrometer, modified by Fluorescence Innovations, Inc. for fluorescence lifetime measurements.

The tunable UV light around 295 nm was generated by pumping a compact dye laser with a pulsed Nd:YAG laser (Teem Photonics, France) and frequency doubling the output. The pump laser is directed into the flow cell, just passing over the right angle prism (Figure 5.16).

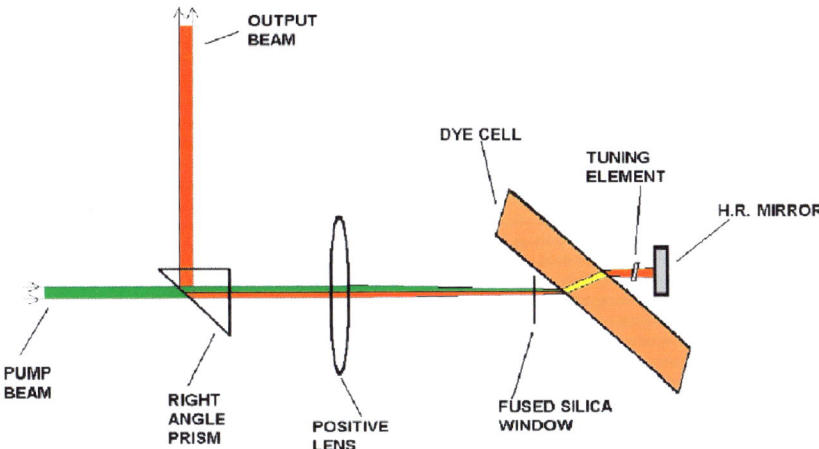

Figure 5.16. Schematic drawing of dye pumped laser configuration. The dye used was Rhadamine 6G pumped by a Nd:YAG source.

A high reflector (H.R.) mirror reflects the output beam back through the tuning element, which can be rotated to select the excitation wavelength. The doubling crystal converts 560-600 nm light to 280-300 nm. Dichroic mirrors, capable of reflecting UV and passing visible light, remove any reflected green light from the UV beam. The UV laser pulse repetition frequency is determined by the 1 kHz Nd:YAG pump source. This is low compared to conventional time domain fluorometers, which operate at MHz frequencies. The high pulse repetition frequencies are not necessary in Fluorescence Innovation's instrument due to implementation of a transient digitizer and lasers with higher pulse energy. The higher pulse energy increases the number of photons emitted and the transient digitizer collects every photon per pulse, improving the photon statistics.

The sample fluorescence is separated using the emission monochromator equipped in the Eclipse (component 6, Figure 5.15). In a standard Eclipse, a mirror redirects the wavelength-selected photons emerging from the exit slit to a Hamamatsu

R928 PMT. However, this PMT is not suitable for collecting *precise* lifetime data on a nanosecond timescale because the PMT transit time is dependent on where the photocathode is irradiated. Therefore, a Hamamatsu R7400 PMT was installed. To retain functionality of the Eclipse, it was mounted on an adjustable slide. For lifetime measurements it sits at the exit slit of the emission monochromator and can be moved to the side for conventional use of the Eclipse.

The signal from the R7400 PMT is sent to FI's patented transient digitizer. The sampling rate of the digitizer is 1 GHz with 5 times interleaving, or effectively 200 picosecond time steps. The entire fluorescence decay curve is measured for every laser pulse, rather than laboriously building up the decay curve one photon event at a time, as is the case with time-correlated single photon counting. This method of digitizing measures lifetime data very rapidly, on the order of 100 times faster than conventional time-correlated single photon counting. Because FI was able to implement lasers with 10,000 times higher energy per pulse, in exchange for lower pulse repetition frequencies, one can easily record the contributions of many thousands of photon events per excitation pulse. This results in better photon statistics and higher precision data.

Data Analysis of Fluorescence Lifetime Measurements

Linear Combination Analysis. By considering the decay spectra as vectors, all data points can be used rather than condensing them to a single value, as done in iterative reconvolution. Experimentally, two vectors describing the fluorescence decay of both free and complexed forms of Con A can be defined (blue and pink curve, respectively, Figure 5.17).

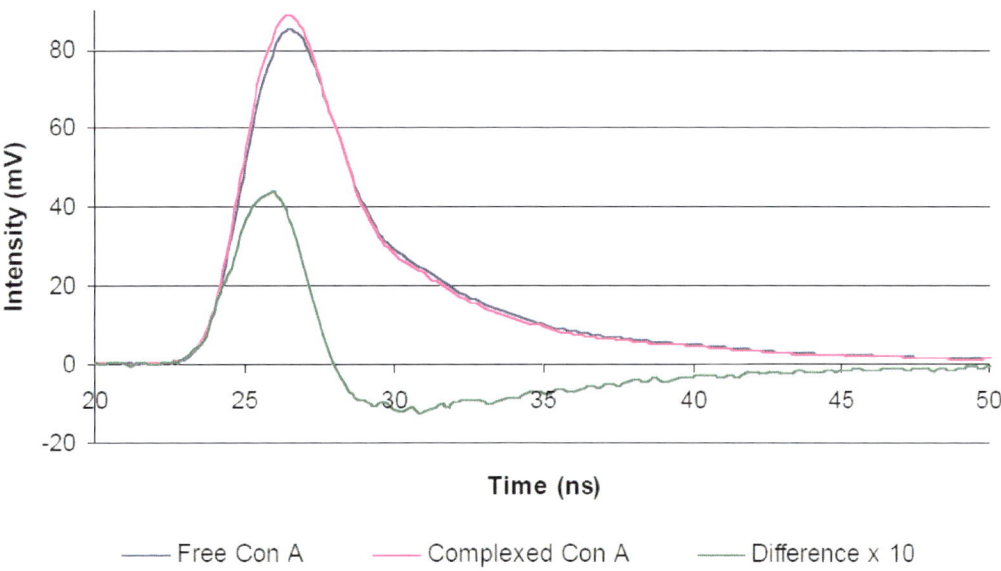

Figure 5.17. Normalized free and complexed waveforms along with their difference scaled by a factor of 10.

The free vector is obtained from the baseline taken preceding glycodendrimer addition and the complexed vector from the last thirty seconds of the most concentrated glycodendrimer addition. A linear combination of these two vectors could be used to fit each decay curve (equation 5.1). Therefore, these vectors are referred to as basis waveforms.

$$W_i = c_{if}W_f + c_{ic}W_c$$

Equation 5.1. Linear combination of basis waveforms.

A waveform at time i, (W_i) can be represented by mixing the basis waveforms of the free and complexed Con A (W_f and W_c respectively). The relative amount of each basis waveform is determined by the preceding coefficients. Hence, for any given time i, c_{if} and c_{ic} identifies how much of the free and complexed waveform, respectively, is present. Consequently, the coefficients reveal the concentration of each species at a given time. As a first approximation, it is assumed that at maximal dendrimer concentration, all Con A has entered into a protein-protein interaction. This waveform is then used as the basis waveform for complexed Con A.

Complexed Waveform Comparison. The complexed decay curves were compared across glycodendrimer generations and shown to be identical (Figure 5.18). This supports the statement that all glycodendrimers, regardless of generation, at saturating concentrations formed similar cross-linked states with Con A.

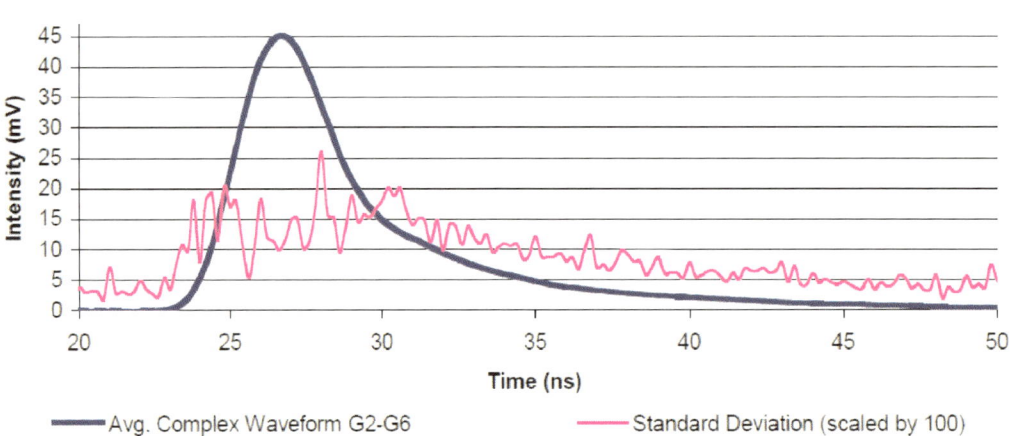

Figure 5.18. Overlap of complexed waveform for G2-G6 and standard deviation scaled by a factor of 100.

Rate Constant Analysis. The coefficients corresponding to complex formation with time were then used to calculate a k_{obs} for the different glycodendrimer generations using equation 5.2.

$$response = response_{max} * (1 - e^{-k_{obs}*(t-t_0)})$$

Equation 5.2. Kinetic association binding for a one-phase association.

Where $response_{max}$ is the maximum conversion for a given glycodendrimer concentration, t is the time in seconds, t_0 is the time of injection, and k_{obs} is the observed binding rate. A non-linear least-squares fit was used to solve for $response_{max}$, k_{obs}, and t_0. The bi-phasic nature of the data can be seen in Figure 5.19. The data was fit to 100 seconds accounting for approximately 90% of the total change. The exceptions were additions of lower concentrations, which required all of the data to be fit (Figure 5.20).

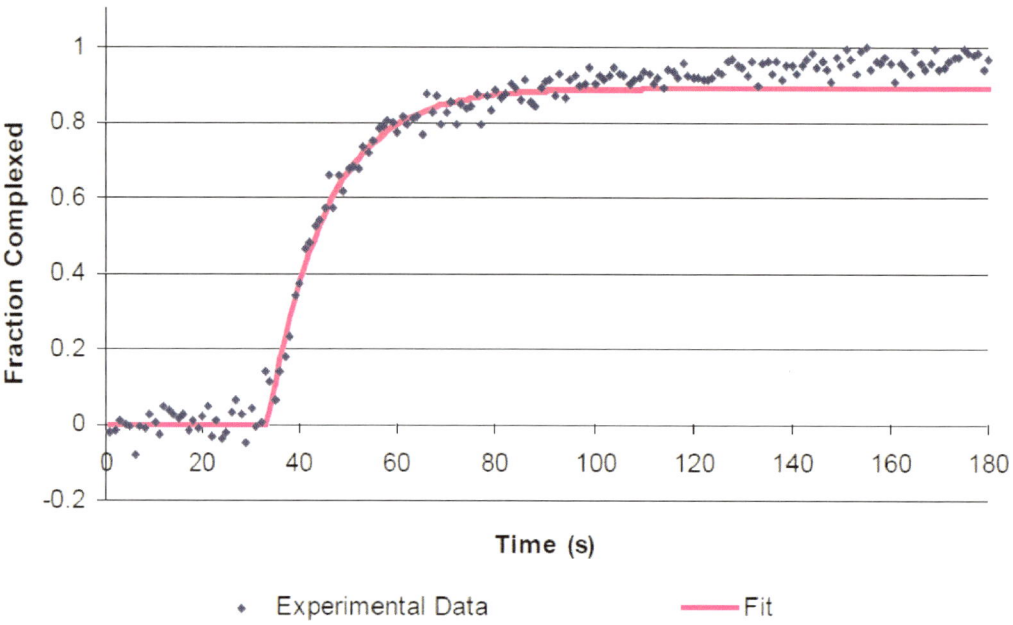

Figure 5.19. Fit of 14.3 μM G(3)-man data for determination of k_{obs}.

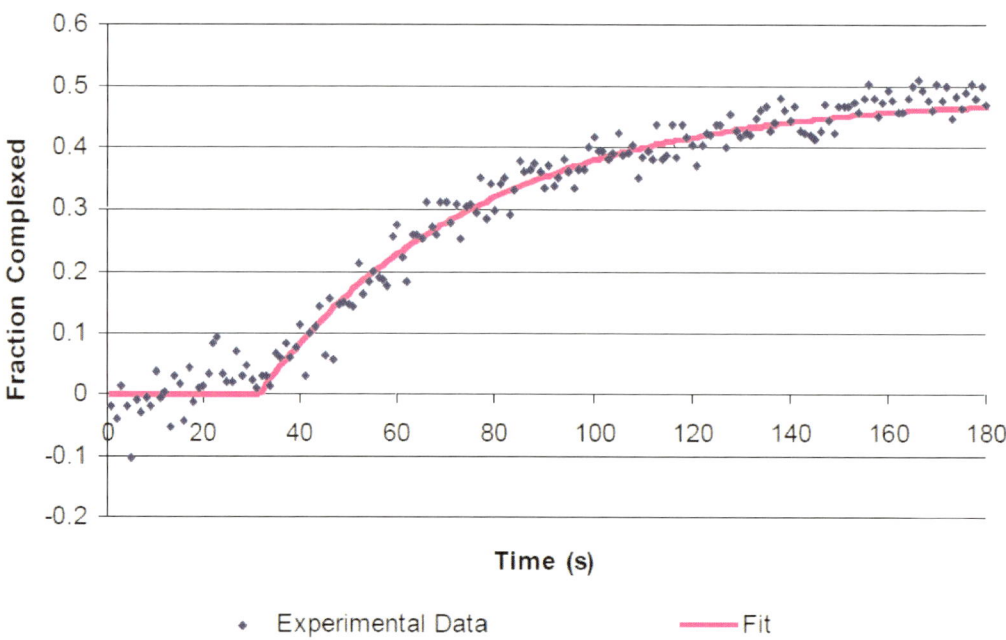

Figure 5.20. Fit of 1.8 μM G(3)-man data for determination of k_{obs}.

By determining the observed rate k_{obs} from an exponential association equation, more precise kinetic rates can be determined if the data is fit globally to obtain values for k_{on} and k_{off}. In this method, k_{on} and k_{off} are constrained to be shared across every kinetic association profile to derive a single best-fit estimate for k_{on} and one for k_{off}.

$$k_{obs} = [dendrimer] * k_{on} + k_{off}$$

Equation 5.3. Observed rate constant expressed as dependent on the rate of association and the rate of dissociation.

By setting the parameters k_{on} and k_{off} to be shared across all the binding experiments for a given compound, the association rate constant and dissociation rate constant can be determined from multiple binding curves instead of only the observed rate constant k_{obs} from individual experiments (Figures 5.21 – 5.24, Table 5.2). However, mannose-functionalized generation 6 dendrimer **6** did not appreciably fit this model, yielding a patterned residual and negative rate constant.

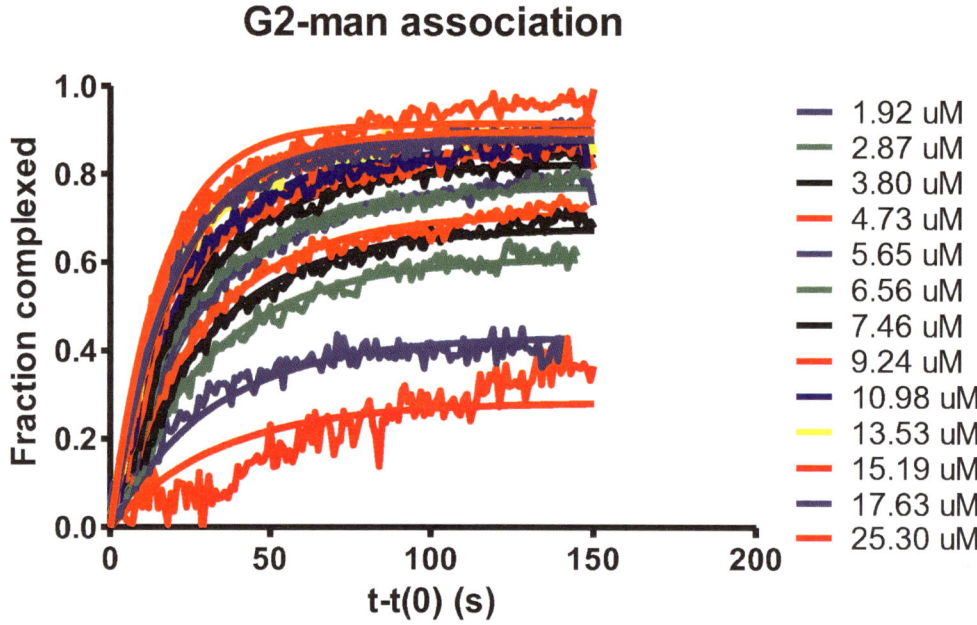

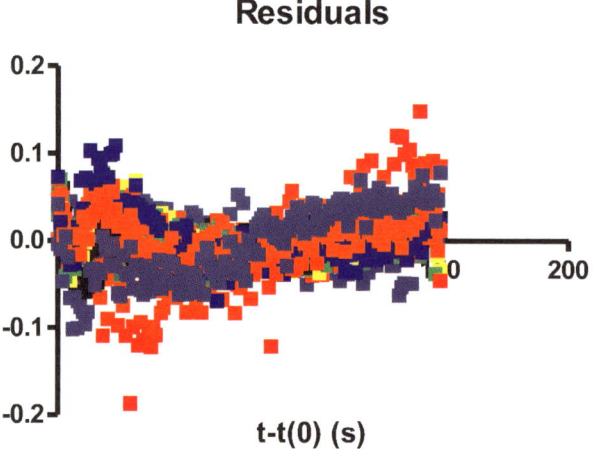

Figure 5.21. Binding data of **2** fit to a 1:1 exponential association model. Bottom: residuals of the non-linear fit.

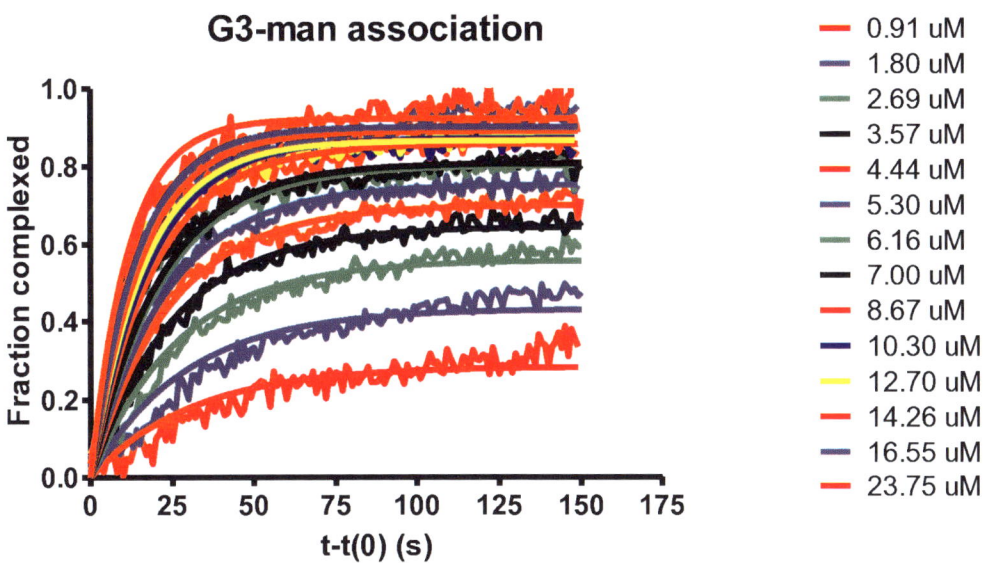

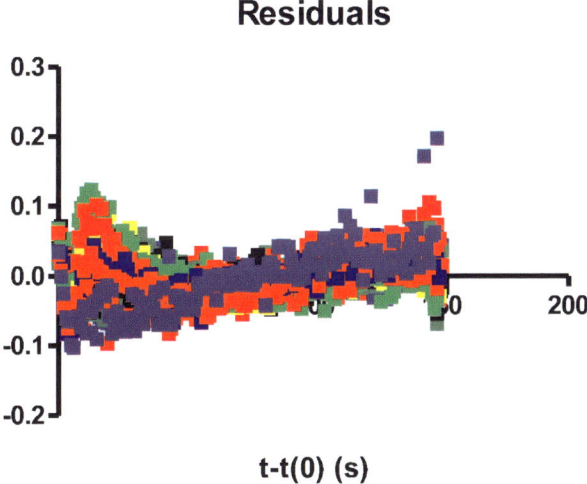

Figure 5.22. Binding data of **3** fit to a 1:1 exponential association model. Bottom: residuals of the non-linear fit.

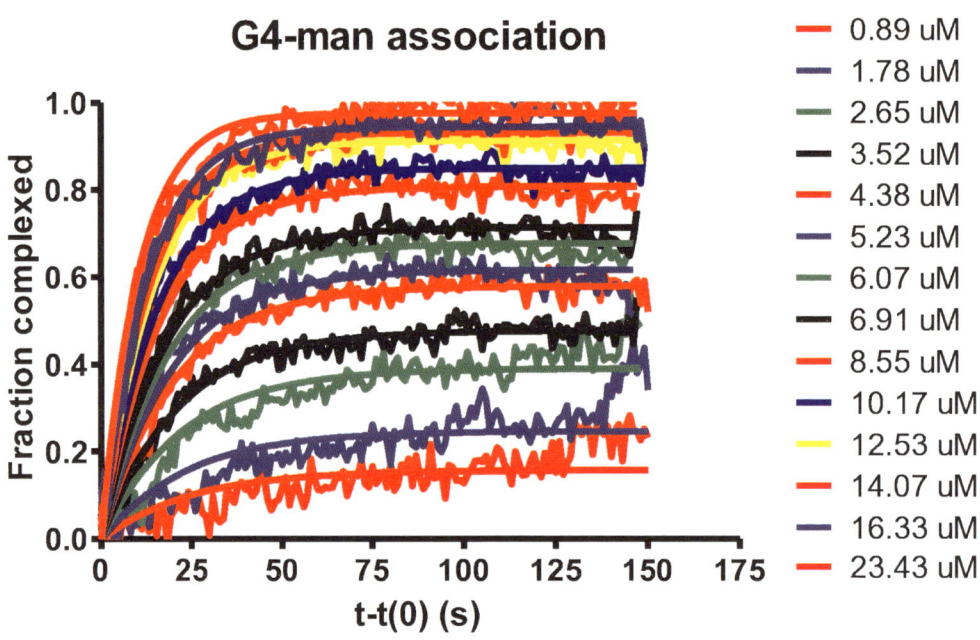

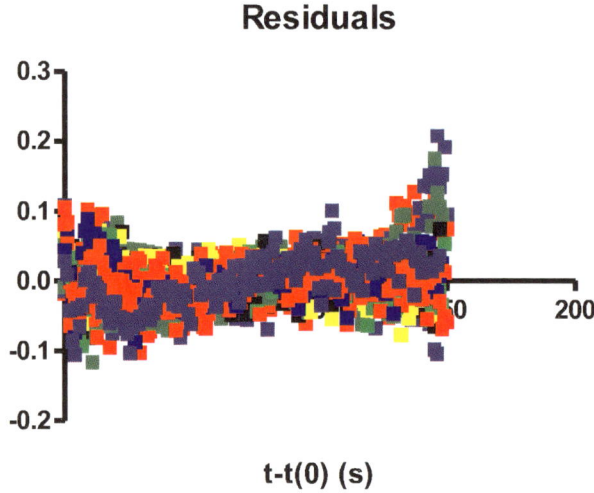

Figure 5.23. Binding data of **4** fit to a 1:1 exponential association model. Bottom: residuals of the non-linear fit.

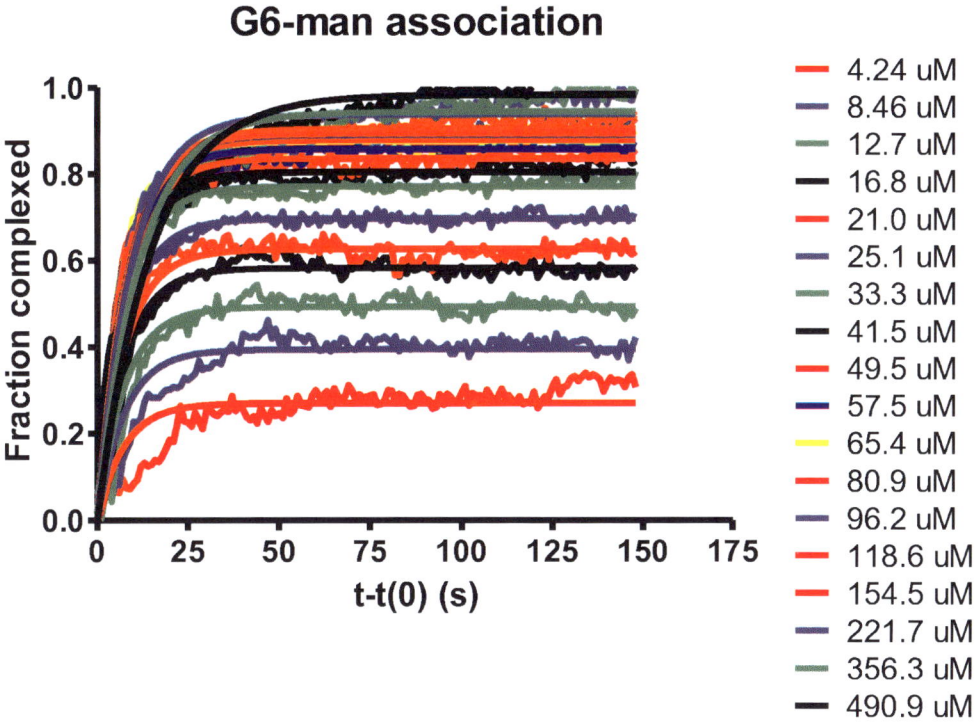
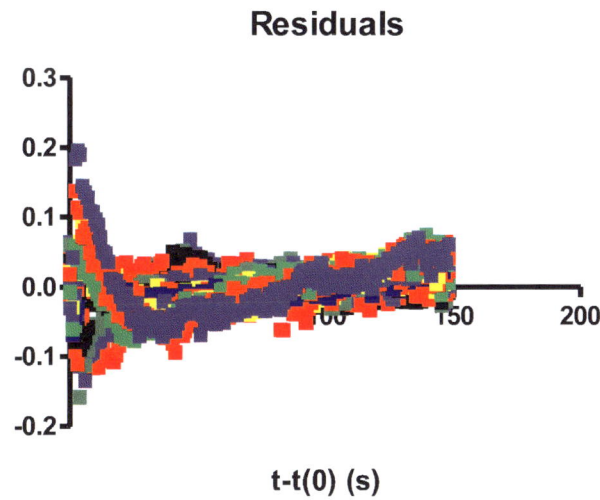

Figure 5.24. Binding data of **6** fit to a 1:1 exponential association model. Bottom: residuals of the non-linear fit.

Experimentals

General Methods. General reagents were purchased from Acros. Concanavalin A (Con A) was purchased from CalBioChem. Glycodendrimers were prepared as preciously described (cf. references 52, 53, 105). A Quartz Fluorometer Cell (3/Q/10-GL14-C) was purchased from Starna Cells, Inc.

Fluorescence Lifetime Acquisitions. Known amounts of glycodendrimer solution were added to 2000 μL of 100 μg/mL Con A in a well-stirred cuvette at 25 °C. The baseline of Con A fluorescence (λ_{ex} = 295 nm, λ_{em} = 335 nm, 5 nm bandpass) was established for 30 seconds before an aliquot of the ligand was added. Fluorescence decay waveforms were measured once per second for the next 130 to 220 seconds.

CHAPTER 6

CONCLUDING REMARKS

Although carbohydrate functionalization is one of the most common post-translational modifications used in biological systems, interactions involving carbohydrates are often weak, and are compensated for by multivalent recognition schemes. The evaluation of these multivalent interactions is the focus of this report, as seen through multiple analytical techniques.

The work described here uses different analytical techniques to evaluate protein-carbohydrate interactions. Many experiments make use of Concanavalin A (Con A), a well-studied protein used previously in our lab. However, the multiantennary nature of the glycodendrimer and the multivalently bound protein combined often result in extensive aggregation, making interpretation of individual binding events impossible. Because of the complicating effects of aggregation of Con A, the first experiments provided in this report make use of pea lectin, a carbohydrate binding protein similar in structure to Con A. Pea lectin is a dimer which binds considerably more weakly to mannose than the tetrameric Con A, so it was hoped that aggregation events could be avoided. Although the aggregation was significantly reduced, the aggregation was still present. However, the nature of the aggregates was different for pea lectin. Although both pea lectin and Con A share bivalent binding motifs with larger glycodendrimers, Con A is able to form a more stable cross-linked lattice and remains in an insoluble aggregate state at varying glycodendrimer concentrations. Pea lectin's aggregates dissociate at non-ideal dendrimer:protein ratios, arising from its less ideal interaction with the glycodendrimer. It was concluded that although Con A and pea lectin share similar structures, shape complementarity plays a significant role in how protein-carbohydrate interactions are expressed on a macromolecular level.

Because pea lectin did not provide us with the reduction in aggregation as hoped, further experiments were done with Con A, the much more readily available and robust alternative. Surface plasmon resonance (SPR) was used to determine the efficacy of a multivalent dendrimer to bind Con A, effectively inhibiting Con A to bind to a self-assembled monolayer of carbohydrates. This has potential applications in drug development, as preventing a targeted protein from reaching the surface of a cell can inhibit infection. Although the kinetics of the system could not be established due to the multivalent nature of interaction, it was determined that regardless of generation, the glycodendrimers have similar inhibition efficacies. These results obtained by SPR were decidedly different than the results obtained by previous hemagglutination assays. This is due to the nature of what SPR measures; in hemagglutination inhibition assays, the glycodendrimers disrupt the ability of red blood cells to form aggregates, whereas SPR provides a more accurate measurement of protein-carbohydrate affinity independent of aggregation and precipitation events.

At this point, it was postulated that a patterned surface would provide an increased affinity more than a random distribution of sugars around the glycodendrimer. ELISA inhibition assays were used to test the efficacy of tris-clustered mannose-functionalized dendrimers. The experiments suggested that tris-clustered mannose residues show no significant difference in inhibition ability over single mannose endgroups when used on dynamic dendrimer scaffolds. Ideal inhibition occurred with smaller dendrimers when compared to larger dendrimers with comparable endgroups. However, trismannose clustering allows for a redistribution of a scaffold's surface functionalities, potentially freeing up endgroups for other functionalities such as imaging or solubility.

From these studies, it became clear that aggregation events induced by glycodendrimers were a prevalent phenomenon, and measuring protein-carbohydrate interactions in the presence of aggregates required a novel approach. Fluorescence Innovations, Inc. had recently developed a fluorescence lifetime reader capable of measuring fluorescence events on the second timescale, independent of aggregation events. Using this technology, the kinetics of glycodendrimer-mediated Con A aggregation could be characterized by fluorescence with high precision.

Although many approaches to elucidate protein-carbohydrate interactions were discussed in this thesis, the field of protein-carbohydrate interactions is massive. This work provides the results of several analytical approaches to studying the field of multivalent protein-carbohydrate interactions. It emphasizes that analysis techniques measuring a single type of interaction, such as a protein-carbohydrate interaction, can yield very different results depending on the experimental setup. The work presented here suggests several methods by which aggregation and cross-linked states can be monitored. This offers a first step to controlling aggregation through the introduction of multivalent compounds. Inhibition of protein-carbohydrate interactions is affected by protein aggregation states and how the interaction is measured greatly affects how the inhibition pathway is understood. Aggregation plays a prevalent role in multivalent affinity enhancement, with many subtle requirements in aggregation formation affecting inhibition of protein-carbohydrate interactions.

APPENDIX A

GLYCOPROTEOMICS WITH BORONIC ACID DERIVATIVES

Introduction

The field of glycoproteomics is a branch of proteomics in which proteins that have been post-translationally glycosylated are characterized. Mass spectrometry is often used in proteomics to identify proteins and their post-translational modifications, but mass spectrometry relies on initial separation techniques such as liquid chromatography or 2D gel electrophoresis.[195] The ability to pinpoint post-translational modifications and to enhance detection sensitivity would dramatically improve the field of glycoproteomics, leading to the development of better diagnostic tools.

Selectively recognizing carbohydrates under physiological conditions has been a difficult challenge in chemical biology.[196] Boronic acids have shown promising results for the recognition of carbohydrates in aqueous solutions[197, 198], but until recently have shown only a limited binding ability to nonreducing sugars, which make up the majority of cell surface glycoconjugates.[199, 200] Hall and co-workers have recently developed a new class of glycoside-binding boronic acids that are able to bind complex glycosides under physiologically relevant conditions.[200] Hall's compounds serve in the design of a new class of receptors to selectively target glycosylated compounds.

Benzophenone photochemistry has been widely used since its first biological application was reported in 1974.[201] Benzophenone has several properties that make it an ideal candidate for inclusion in covalent modifications.[202] Benzophenones can be manipulated in ambient light and are not activated until exposed to 350-360 nm light, which also avoids protein damaging wavelengths. Benzophenones react preferentially with unreactive C-H bonds even in the presence of solvent water and bulk nucleophiles. These advantages are favorable enough to outweigh the disadvantages such as the additional bulk and hydrophobicity that are caused by the introduction of the benzophenone group.

Irradiation of the benzophenone results in the promotion of an electron in the carbonyl to an excited triplet state, which allows for interactions with a weak C-H bond, leading to hydrogen abstraction and C-C bond formation.[202] If a suitable H-donor with the required geometry is not found, the triplet state quickly relaxes to the ground state, maintaining the photoactivation properties until a suitable geometry is found. This reversibility makes benzophenone an ideal candidate for interactions in aqueous media. To increase the probability of successful C-H bond insertion, a flexible linker to the benzophenone from the binding component can be inserted to improve the coordination necessary for reaction.

Fluorescent dyes have generally been limited by their often hydrophobic structure. The Grieco group has made advances in creating Z-dyes, a family of highly water soluble, ultrasensitive fluorescent, zwitterionic multiplex proteomic detection dyes that have been designed as an alternative to traditional hydrophobic dyes.[203] The enhanced solubility of the Z-dyes provides an increased sensitivity for protein detection, especially for detection of difficult proteins such as membrane proteins. Z-dyes also provide improved quantification of protein levels, and are particularly useful when detecting changes due to post-translational modifications arising from biological stimuli.

ARS Assays

Combining the boronic acid and benzophenone moieties allows for the development of a glycosensor **21** with enhanced sensitivity (Figure A.1).[204] The boronic acid is able to bind carbohydrates, which places the benzophenone in close proximity to the glycosidic ring. The amino group allows for future functionalization with prospective Z-dyes.

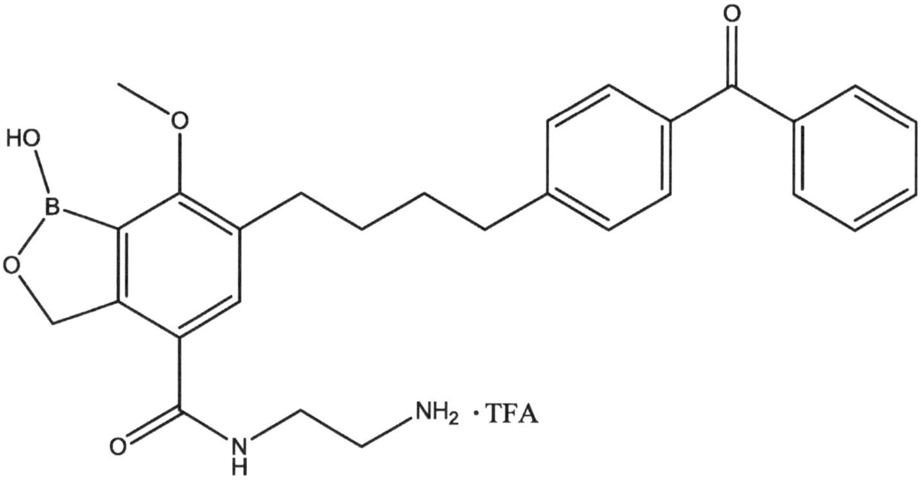

Figure A.1. Glycosensor **21** based on boronic acid and benzophenone.

The viability of **21** for use as a glycosensor was at first tested by determining its ability to bind carbohydrates in aqueous solutions. Initial experiments were performed using phenylboronic acid and *ortho*-hydroxymethyl phenylboronic acid as model compounds to determine the suitability of an assay, as synthesis of **21** is difficult.

For the Alizarin Red S (ARS) assay, both absorbance measurements and fluorescence measurements are viable methods to track binding to ARS. However, absorbance measurements proved to be more rapid and stable than fluorescence measurements. The assay was optimized with phenylboronic acid (PBA) as a model boronic acid and fructose as an initial model sugar, similar to the three-component assay described by Springsteen and Wang[205]. Because subsequent experiments required the judicious use of custom boronic acids, the experiments were modified for minimal use of PBA.

For these experiments, a solution of 0.1 mM ARS was made in a 0.1 M phosphate buffer solution, pH 7.4. First, the association constant was measured for the PBA-ARS complex by determining the absorbance at 450 nm for 100 uL of 5 mM PBA dissolved in 0.1 mM ARS solution. This solution was then diluted with additions of 0.1 mM ARS to a total volume of 2 mL, monitoring the change in absorbance. The association constant was determined as previously described[205] to yield a K_a of 1800 M^{-1}, compared to 1300 M^{-1} obtained by Dr. Wang's group[205] (Figure A.2). The concentration of PBA yielding a ~80% response was determined for use in the three-component competitive ARS assay (2 mM PBA dissolved in 0.1 mM ARS).

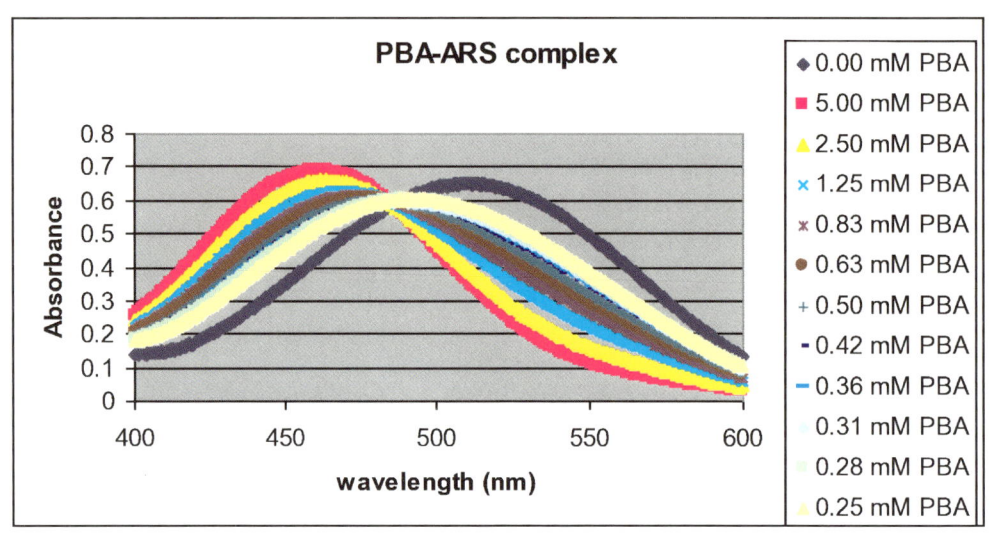

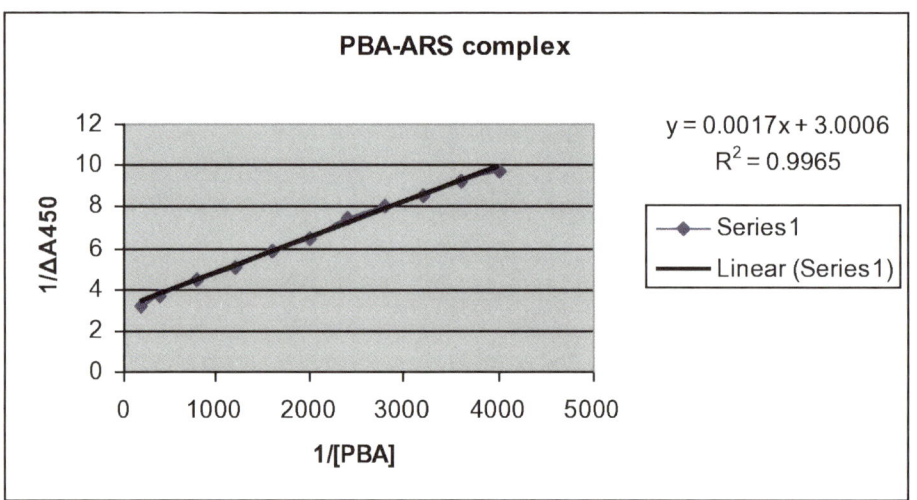

Figure A.2. Absorbance profile of phenylboronic acid (PBA) binding to Alizarin Red S (ARS). Bottom: double reciprocal plot from which K_{eq} for the PBA-ARS interaction is determined.

The association constants for the boronic acid-sugar complexes were found by titrating the PBA-ARS solution with a carbohydrate. Fructose was used as a model sugar for optimization of the inhibition procedure. To this end, a 0.2 M fructose solution was prepared in the 2 mM PBA, 0.1 mM ARS solution, and added to a solution of 2 mM PBA in 0.1 mM ARS, yielding a range of fructose concentrations (15-89 mM). The absorbances were measured at 450 nm and the association constant was determined as previously described[205] to yield a K_{eq} of 177 M^{-1} compared to 160 M^{-1} obtained by Dr. Wang's group (Figure A.3).

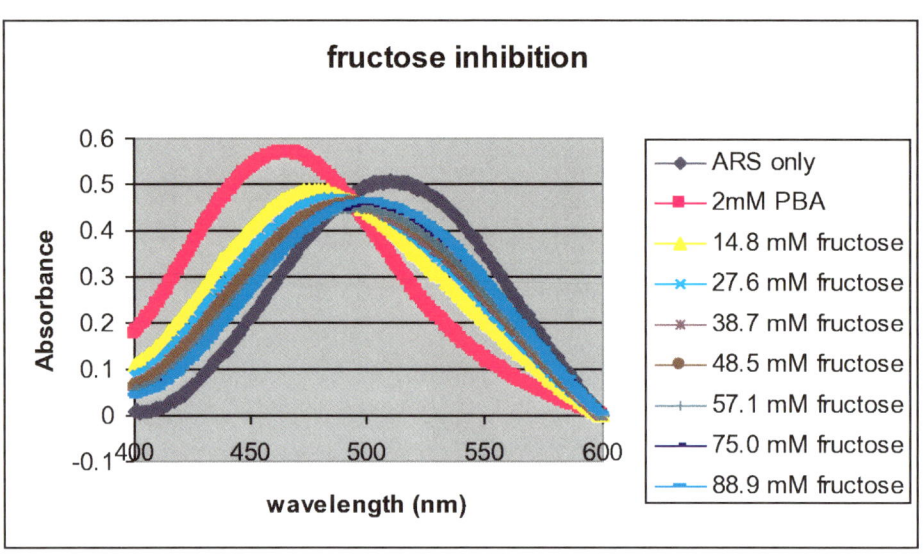

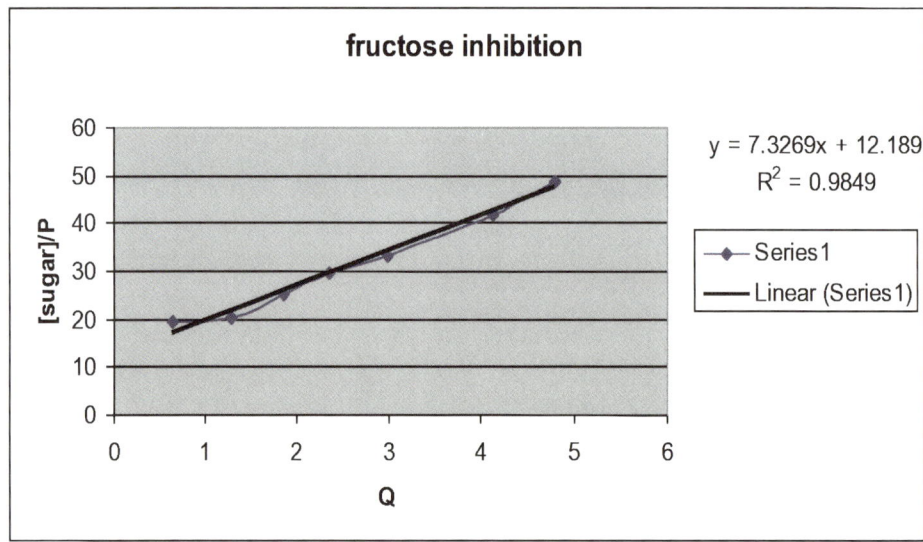

Figure A.3. Absorbance profile of phenylboronic acid (PBA) binding to Alizarin Red S (ARS), inhibited by fructose additions. Bottom: plot from which K_{eq} for the PBA-fructose interaction is determined.

To ensure compatibility with other boronic acids, the assay was repeated using *o*-hydroxymethyl phenylboronic acid to yield association constants of 1275 M^{-1} with ARS,

280 M^{-1} with fructose, and 12 M^{-1} with α-methyl mannose (Figure A.4). This is comparable to values obtained by Dowlut and Hall in their previous studies[200].

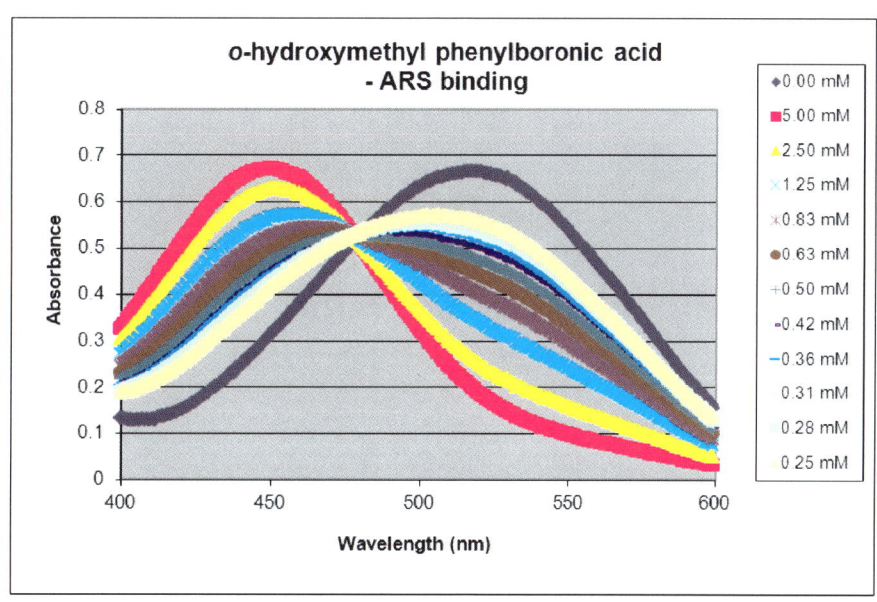

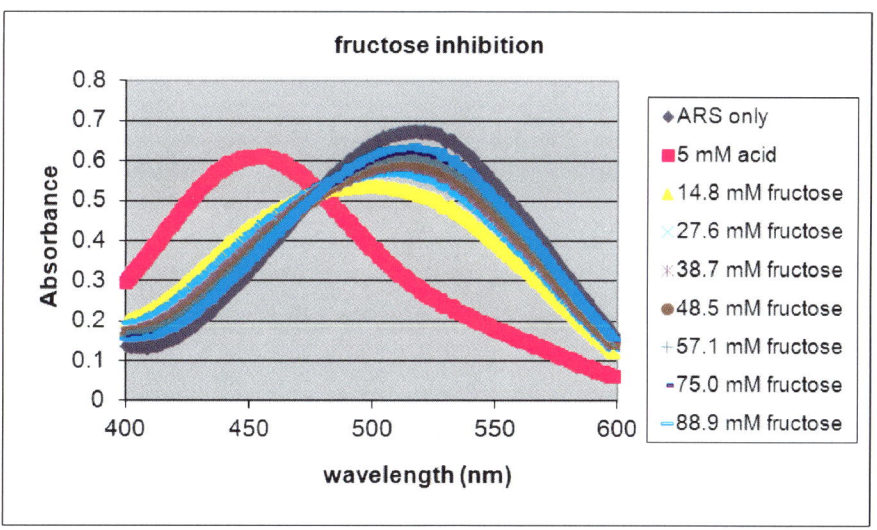

Figure A.4. Top: Absorbance profile of *o*-hydroxymethyl phenyl boronic acid binding to Alizarin Red S (ARS). Bottom: Inhibition by fructose additions.

Preliminary assays using phenyl boronic acid (PBA) and *o*-hydroxymethyl phenylboronic acid were very promising, resulting in association constants comparable to previously published results. As a precursor to **21**, the ARS assay was attempted with compound **22** (Figure A.5). However, **22** failed to elicit any noticeable change in absorbance upon addition of ARS. Because the ARS assay is dependent on noticeable changes in absorbance, this approach was abandoned.

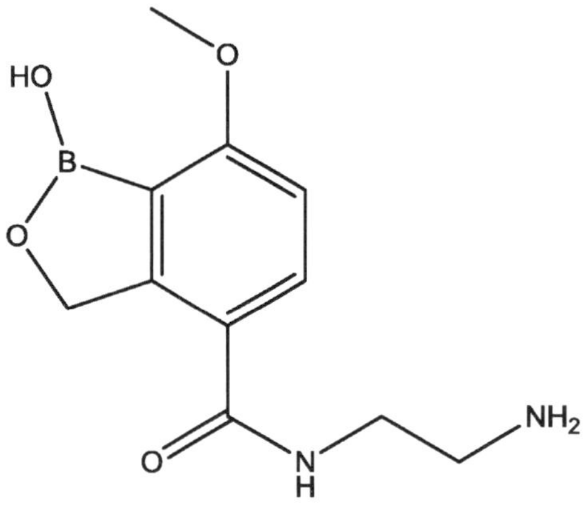

22

Figure A.5. Precursor glycosensor **22**.

NMR Assays

As an alternative, an assay based on ^{1}H-NMR was attempted as described by Dowlut and Hall[200]. Titration of a carbohydrate leads to a distinct chemical shifts of protons on the aromatic ring adjacent to the boronic acid moiety. As a control, fructose additions were made to *o*-hydroxymethyl phenylboronic acid and followed by ^{1}H-NMR

(Figure A.6, Figure A.7). An association constant K_a was obtained by plotting the ratio of peaks found from the ^{1}H-NMR spectrum, consistent with results obtained by Dowlut and Hall (Table A.1, Figure A.8). However, the protons on the aromatic ring adjacent to the boronic acid moiety are not existent on **22**, and further experimentation using this approach was not preferred, as the changes in chemical shifts were too small to be measured when the carbohydrate was added (Figure A.9, Figure A.10).

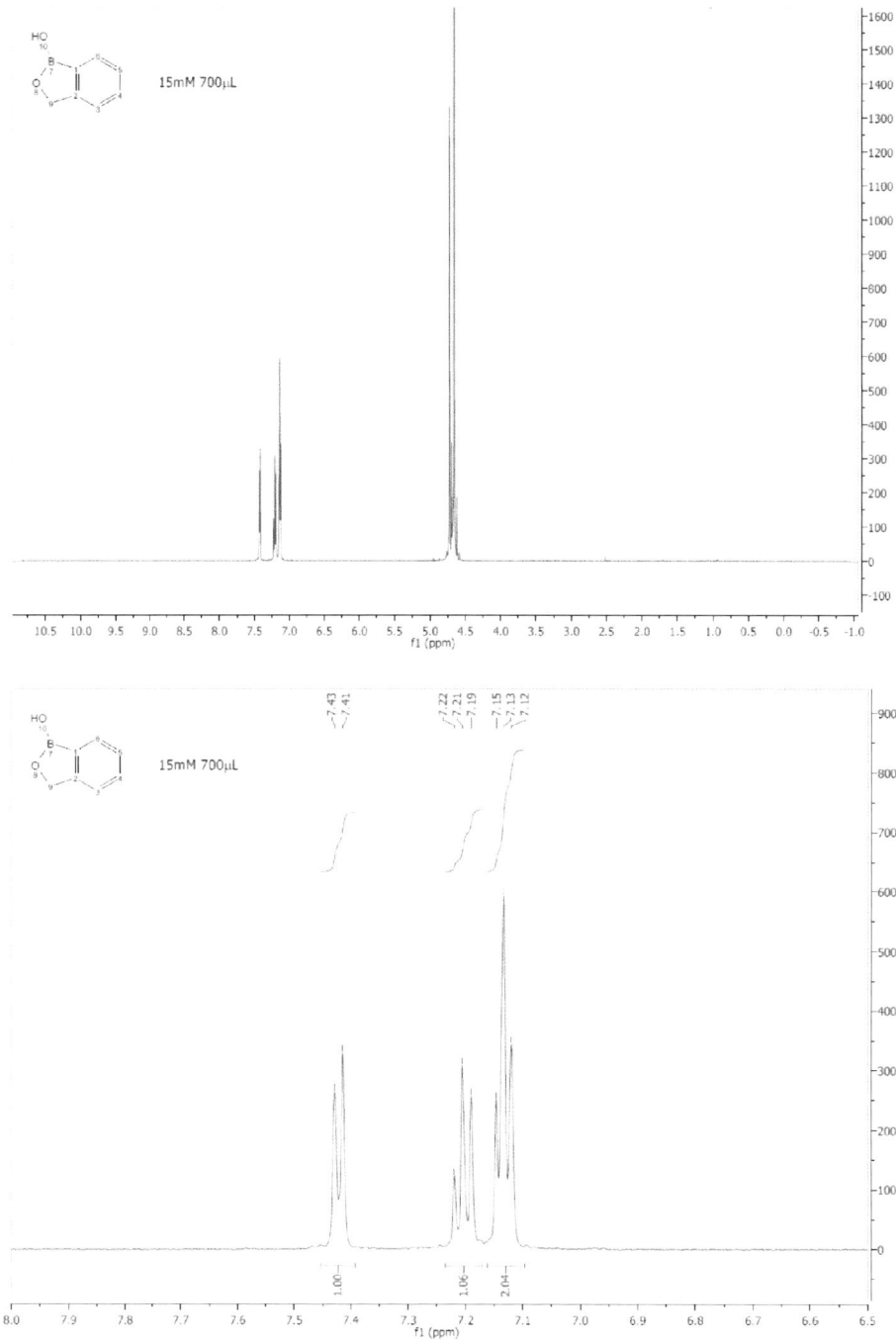

Figure A.6. ^{1}H-NMR of *o*-hydroxymethyl phenylboronic acid before addition of fructose. Bottom: expanded view.

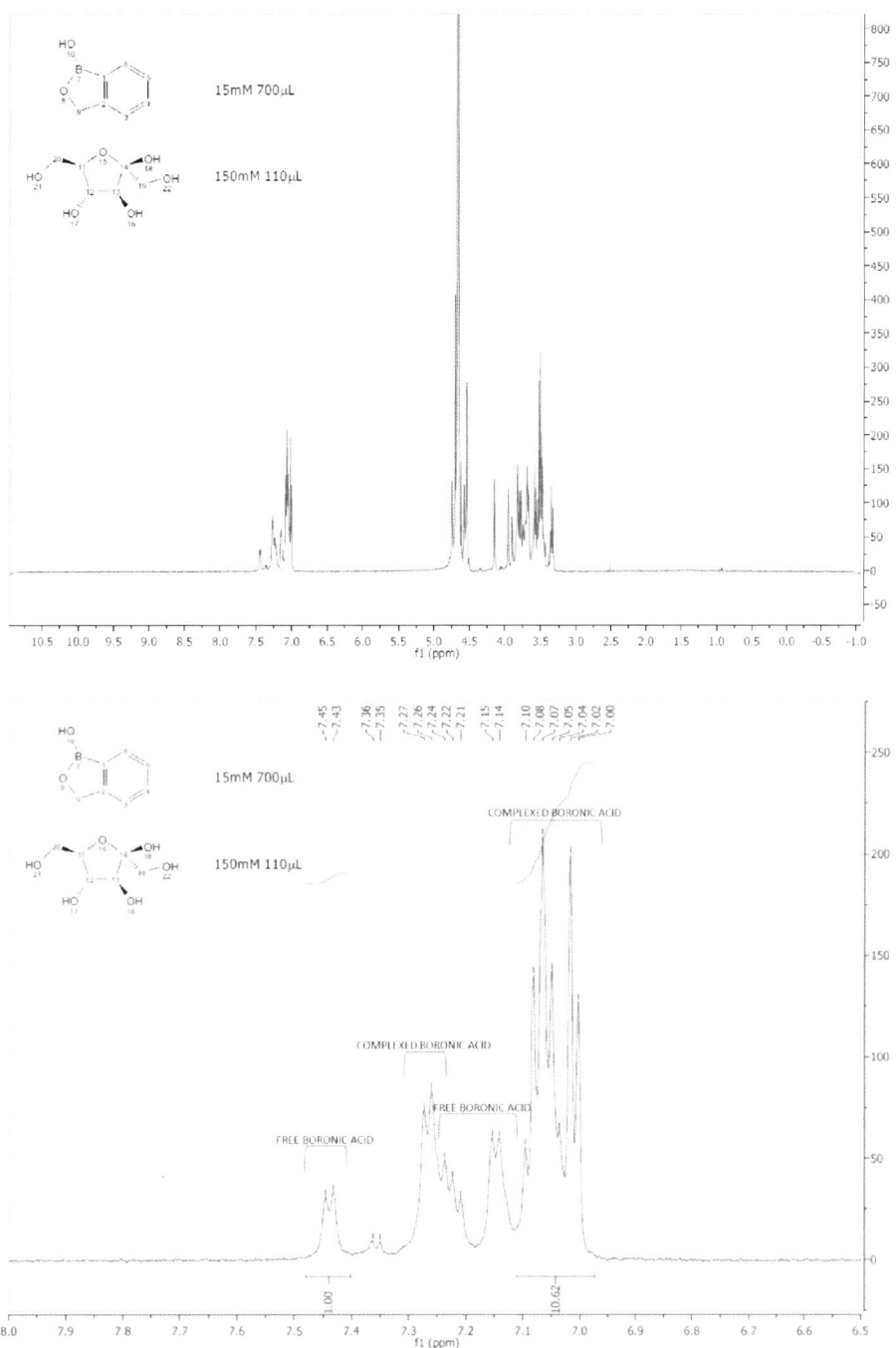

Figure A.7. ^{1}H-NMR of *o*-hydroxymethyl phenylboronic acid after addition of fructose. Bottom: expanded view. The ratio of complexed to free boronic acid can be determined by the ratio of the integrals.

Table A.1. ¹H-NMR titrations of **22** (15 mM) with fructose (150 mM). Equations are described in the experimental section.

exp #	μL fructose	Volume (μL)	[fructose] (M)	[RS]/[R]	1/[S] = 1/[fructose]	1/θ
1	0	700	0.000	0		
2	40	740	0.008	1.65	123.33	1.61
3	50	750	0.010	3.51	100.00	1.29
4	60	760	0.012	4.47	84.44	1.22
5	70	770	0.014	5.22	73.33	1.19
6	80	780	0.015	5.71	65.00	1.18
7	90	790	0.017	7.63	58.52	1.13
8	100	800	0.019	9.19	53.33	1.11
9	110	810	0.020	10.62	49.09	1.09

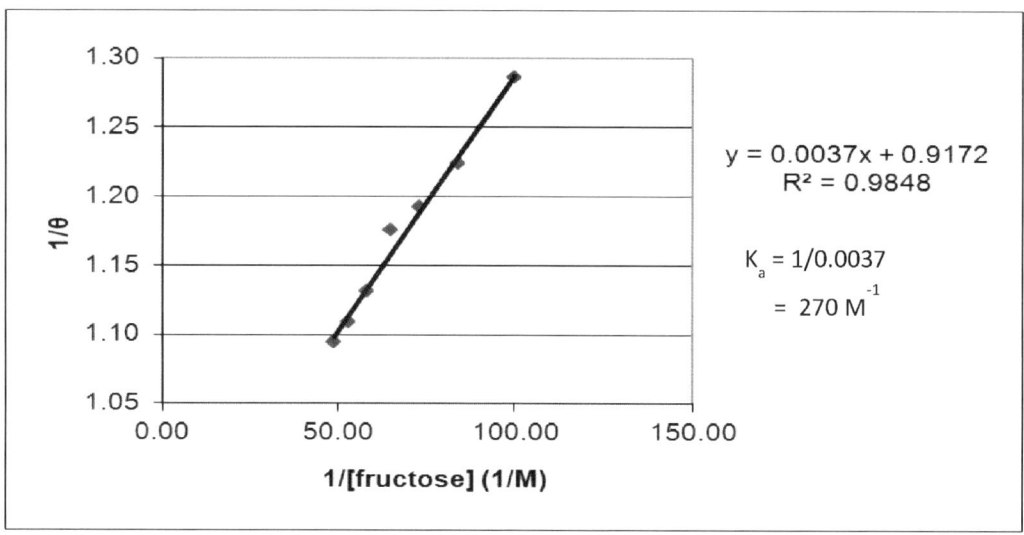

Figure A.8. Double reciprocal plot to determine the K_a of the boronic acid-fructose interaction.

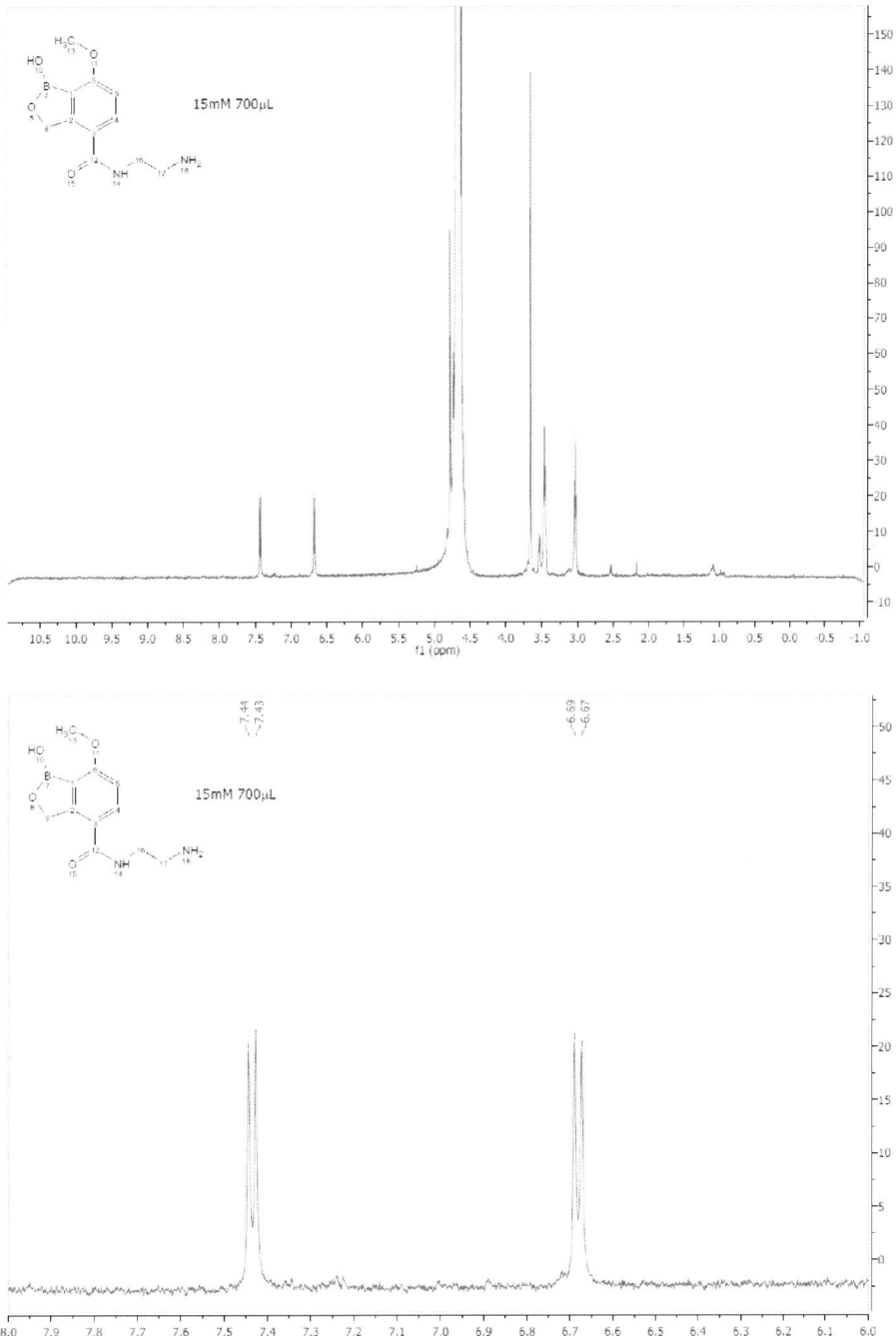

Figure A.9. ^{1}H-NMR of **22** before addition of fructose. Bottom: expanded view.

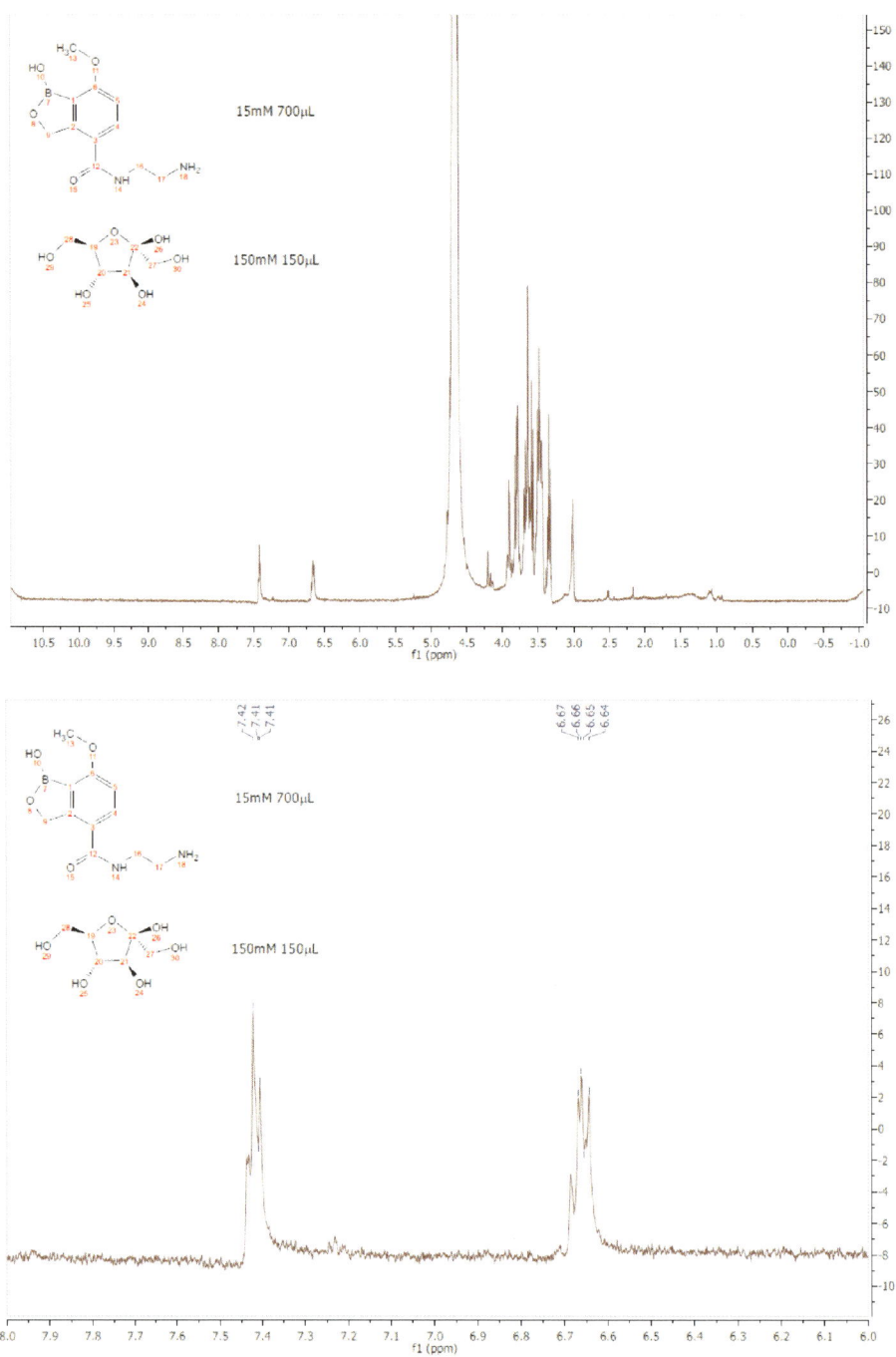

Figure A.10. ^{1}H-NMR of **22** after addition of fructose. Bottom: expanded view. The signals corresponding to the complexed and free boronic acid are significantly overlapped.

ITC Assays

Isothermal titration calorimetry (ITC) requires significantly more material to obtain acceptable data but allows for the determination of thermodynamic data by measuring the change in heat evolved from binding. ITC of **22** yielded good binding data with fructose and sialic acid, but did not result in appreciable heats of enthalpy upon binding lower affinity carbohydrates such as *O*-methyl-*N*-acetylglucosamine (Me-glcNAc) (Figure A.10). For this assay, competitive ITC was used as described by Zhang.[206] The competitive assay was accomplished by titrating fructose, a strong binder, into a solution of **22** and Me-glcNAc (Figure A.11). The resulting equilibrium constant was then compared to the uninhibited equilibrium.

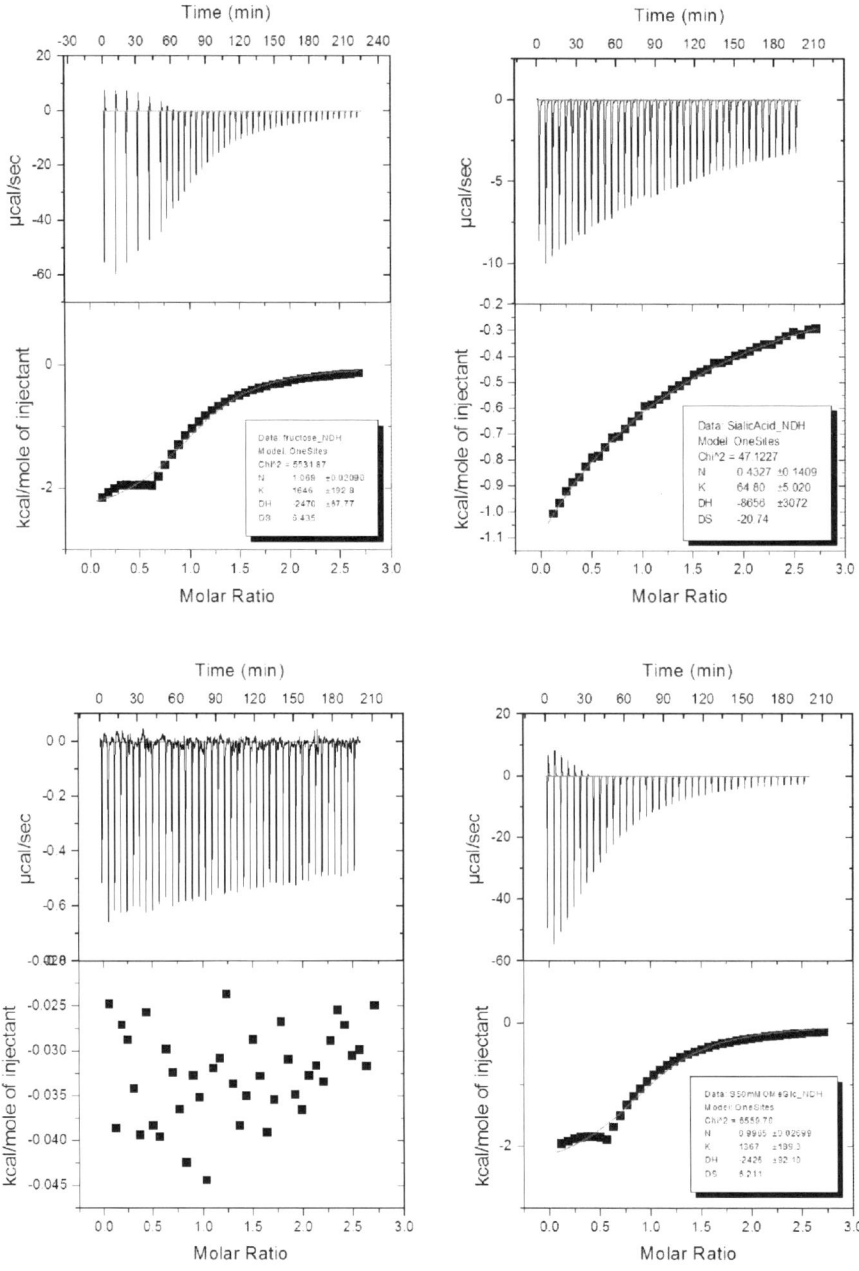

Figure A.11. ITC profiles of a) 75 mM fructose, b) 75 mM sialic acid, and c) 75 mM Me-glcNAc injected into 5 mM boronic acid in 0.1 M PBS, pH 7.4. d) competitive ITC profile of 75 mM fructose injected into 5 mM boronic acid and 50 mM Me-glcNAc.

Titration of **22** with the mentioned carbohydrates yielded a 1:1 binding stoichiometry as expected, but did not give a consistent stoichiometry when titrated with sialic acid. Between the repititions performed with sialic acid, the binding stoichiometry was not reasonable. For this reason, the binding stoichiometry was constrained to N=1, representing the expected ligand:receptor ratio. Furthermore, the additions yielding endothermic spikes were removed to improve curve fitting (Figure A.12). These endothermic spikes are likely the result of slight fluctuations in pH arising from carbohydrate additions into the adiabatic cell. These heats of dilution could be subtracted from future experiments by repeating the carbohydrate injections into a cell containing only the sample buffer. However, since the goal of this experiment was to determine the overall binding ability of **22** to select carbohydrates, this was not deemed necessary in the interest of developing the glycosensor.

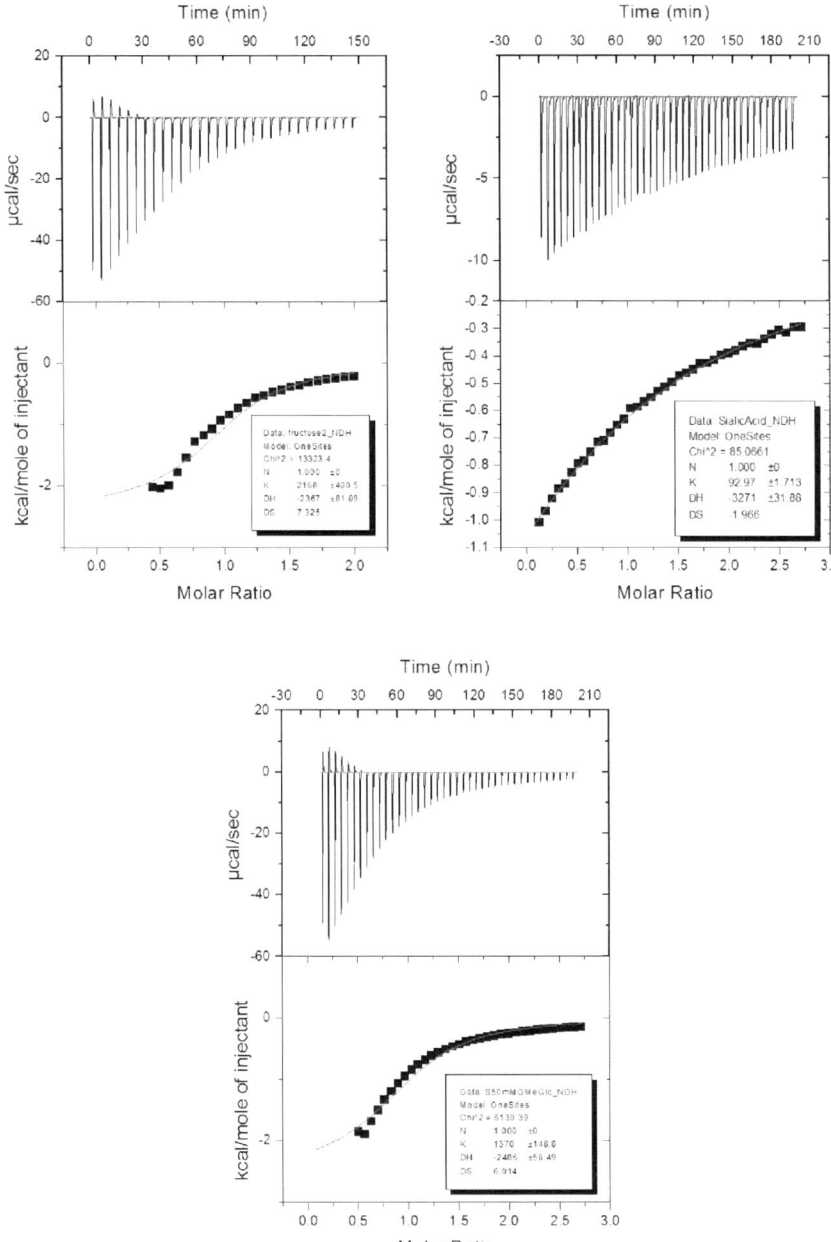

Figure A.12. ITC profiles after removing injections resulting from endothermic spikes and constraining N=1.

From the ITC experiments, compound **22** was found to bind to fructose with an association constant of 1790 M^{-1}, which is comparable to the published binding constant of 610 M^{-1} for hydroxymethyl phenylboronic acid[200]. *O*-methyl-*N*-acetylglycosamine binds with an association constant of 10 M^{-1}, and sialic acid binds with an association constant of 90 M^{-1}. These values were obtained from experiments performed in duplicate using isothermal titration microcalorimetry, either directly (fructose and sialic acid) or in an inhibition binding format (*O*-methyl glcNAc).

Mass Spectrometry of Covalently Attached Glycosensors

Confirming the ability of glycosensor **22** to appreciably bind carbohydrates allowed us to test the ability of glycosensor **21** to covalently cross-link to glycoproteins after selectively targeting them. A 50 μM solution of **21** was made in water and mixed with a solution of fetuin. Fetuin is a protein glycosylated with sialic acid and a reported molecular weight of 48.4 kDa.[207] In theory, the boronic acid moiety of **21** is able to bind the sialic acid of fetuin long enough for the benzophenone moiety of **21** to come into close proximity of the sialic acid, such that a covalent bond can be made upon irradiation.[202]

Cross-linking attempts of **21** to fetuin did not yield conclusive results. Fetuin contains several sialic acid residues added as post-translational modifications, but not all glycosylation sites are always successfully modified. Therefore, any shift due to cross-linking of **21** is more difficult to discern from omissions in sialation sites (Figure A.13).

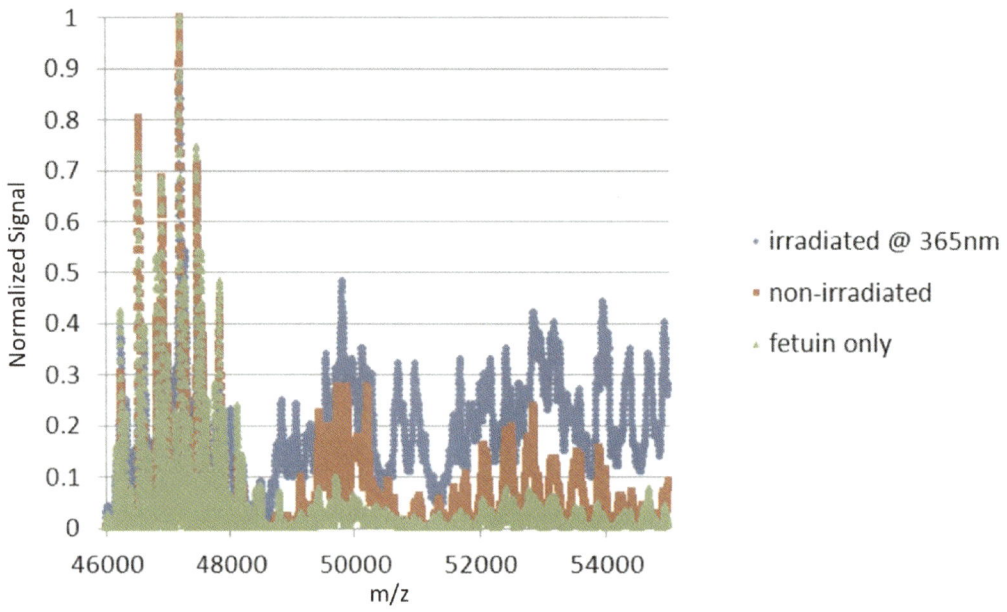

Figure A.13. Mass spectrometry of fetuin and **21**.

To simplify the experiment, simple carbohydrates were used in lieu of the glycoprotein. In these experiments a 1:1 ratio of carbohydrate:glycosensor were combined then irradiated to attempt cross-linking. The samples were then submitted for analysis by mass spectroscopy. While initial results looked promising, they were not consistent in later experiments, possibly suggesting the compounds degrade with time. Early experiments were conducted in water, cross-linking *N*-Acetyl glucosamine to **21** as described above. Additions of *N*-Acetyl glucosamine can be seen adding to **21**, consistent with a molecular weight shift of 235 and a total weight of 723 (Figure A.14). However, it is possible that irradiation causes **21** to degrade significantly, as a peak at the projected 487 is significantly diminished as well as the product peak at 723. Multiple additions of *N*-Acetyl glucosamine can be seen where only a single addition would have been expected. Furthermore, these additions were also seen in the control experiment not irradiated at 365 nm (Figure A.15).

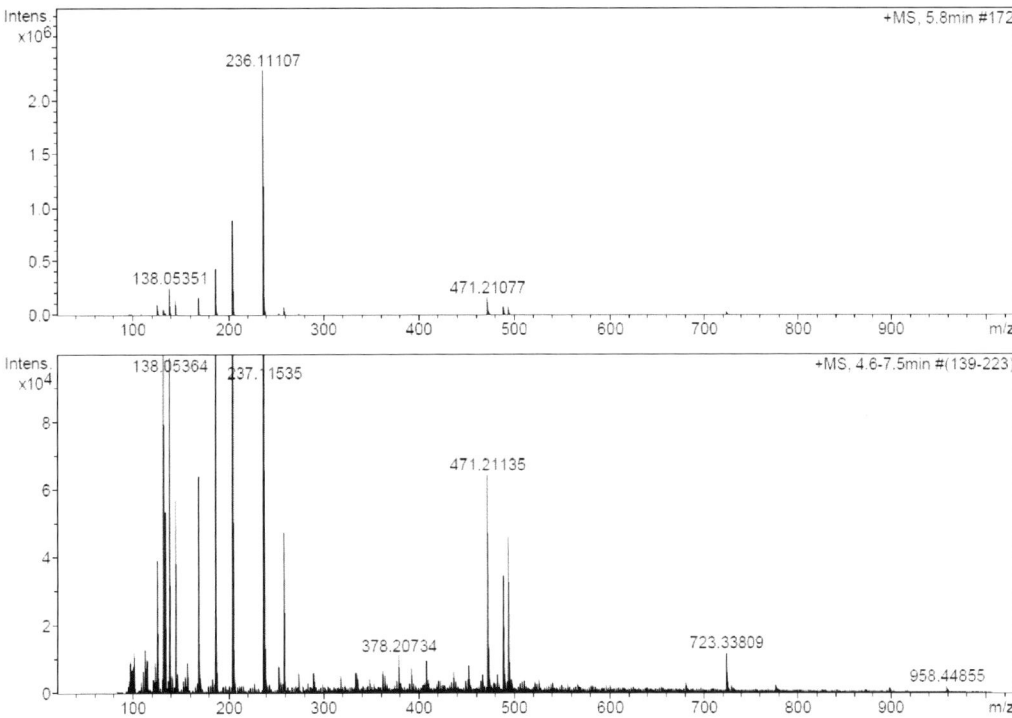

Figure A.14. Mass spectrometry of *N*-Acetyl glucosamine and **21**, irradiated at 365 nm in water. Bottom: expanded view.

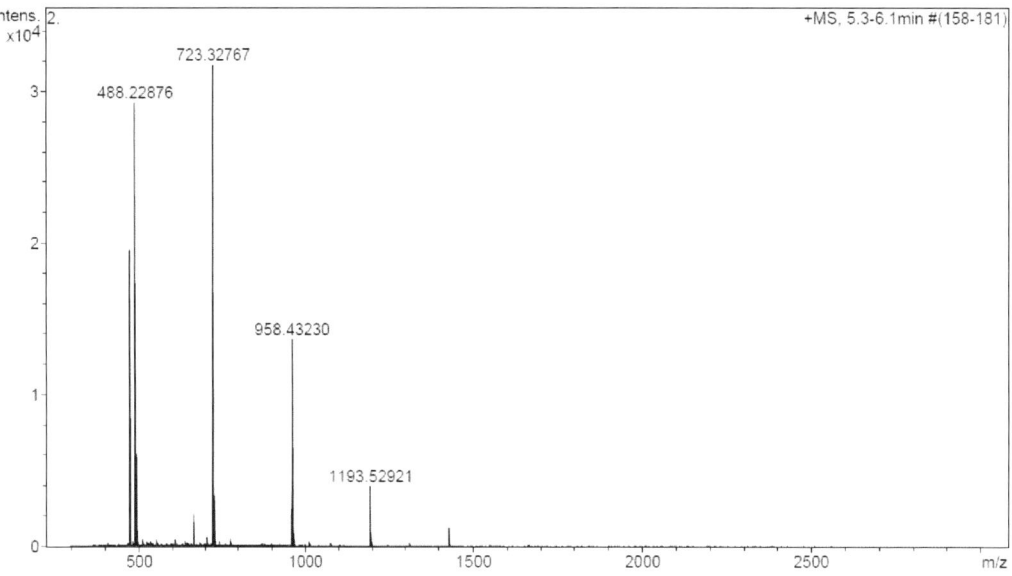

Figure A.15. Mass spectrometry of *N*-Acetyl glucosamine and **21**, control experiment in water.

The experiment was repeated under several conditions, using PBS as a buffer and to increase ionic strength, as well as perfluoro-tert-butanol to decrease ionic strength. Under these conditions, no addition could be observed, and **21** could not be found. Repeating the experiment in water at a later date could not replicate the previous result and did not yield the correct mass for **21**, suggesting that the sample had degraded over time. This was confirmed by ^{1}H-NMR.

Using a fresh batch of material, the experiment was repeated once more in water. The synthesis of the compound had progressed to the point that an Alexa 555 fluorescent dye was added, yielding compound **23** and bringing the synthetic product closer to its intended goal. Because the structure of Alexa 555 is proprietary to Life Technologies, Inc., the exact structure of **23** is not given (Figure A.16).

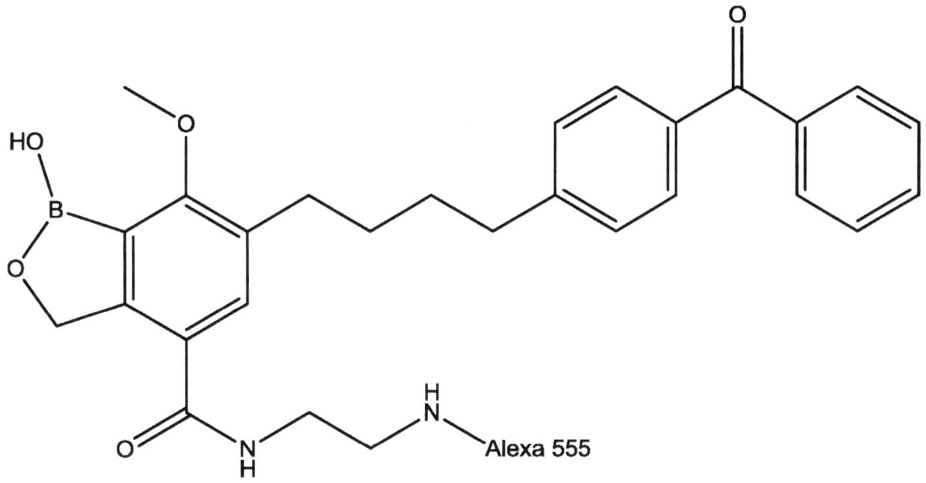

23

Figure A.16. Alexa 555, benzophenone conjugated glycosensor **23**.

Mass spectrometry by electrospray was unsuccessful with **23**, and matrix assisted laser desorption/ionization (MALDI) was used instead. However, MALDI did not yield expected molecular weights for either the cross-linking or control experiments (Figure A.17, Figure A.18). The most noticeable difference is the disappearance of two smaller compounds with molecular weights of 829 and 1057 upon irradiation. MALDI is a soft ionization technique that generally prevents fragmentation, and these peaks do not correspond to the expected molecular weight of **23**. After irradiation these peaks are not present, suggesting that irradiation had possibly caused the compounds attributed to the smaller peaks to degrade, as new peaks of larger molecular weights were not observed after irradiation.

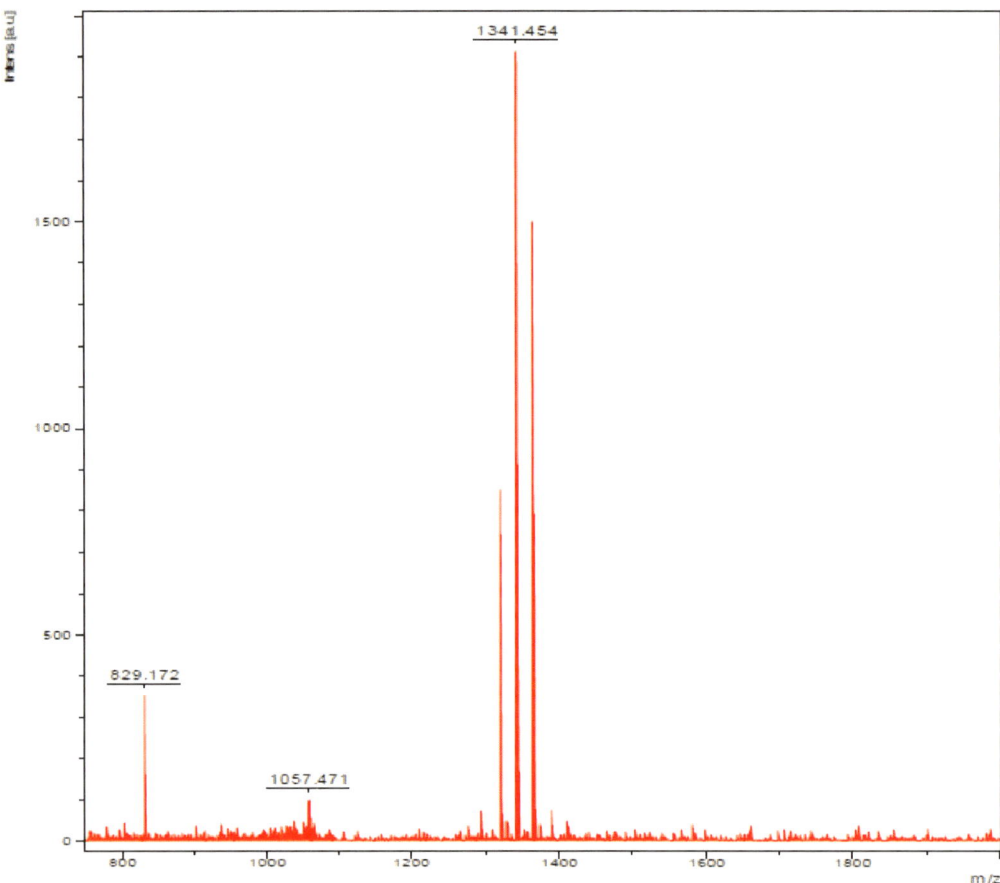

Figure A.17. Mass spectrometry of *N*-Acetyl glucosamine and **23**, control experiment in water.

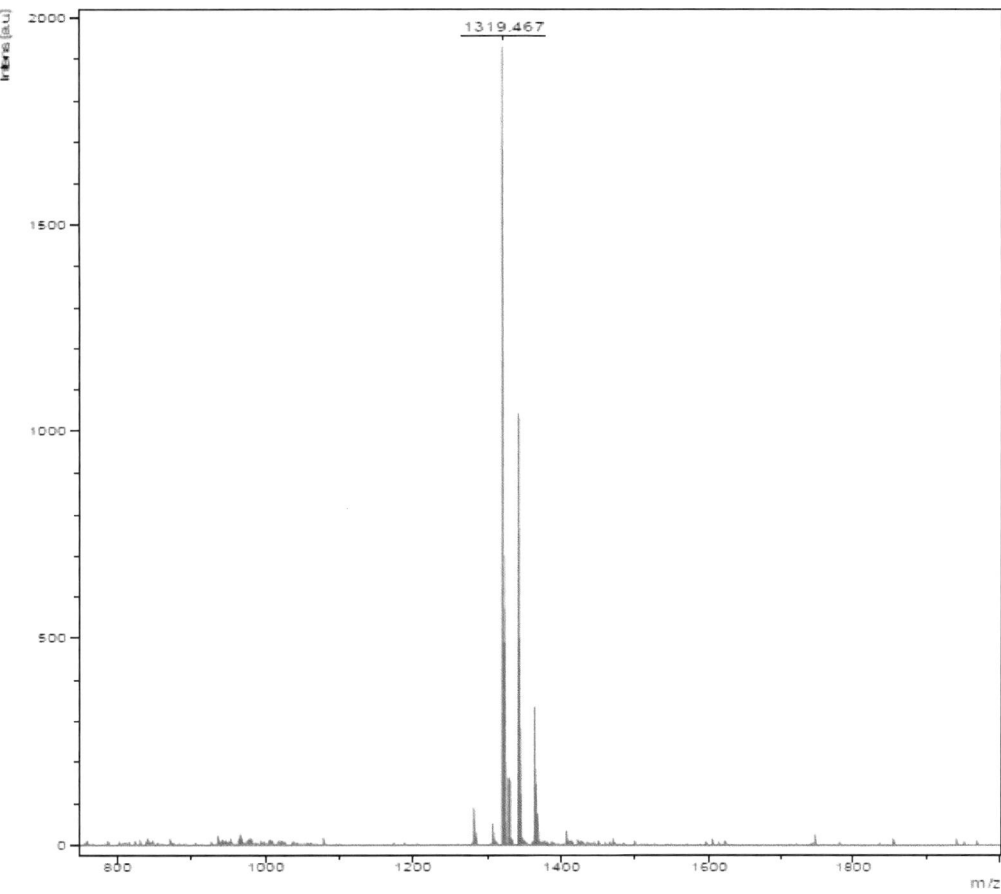

Figure A.18. Mass spectrometry of *N*-Acetyl glucosamine and **23**, irradiated at 365 nm in water.

The peaks around 1300-1400 are present both before and after the irradiation process. The Alexa 555 -conjugated glycosensor has a molecular weight of 1086 determined by mass spectrometry[208], whereas *N*-Acetyl glucosamine has a reported molecular weight of 235, which makes the peak at 1319 a probable candidate for the addition product. As the peak is present in large quantities both before and after irradiation, the carbohydrate addition is likely the result of non-covalent interactions occurring, such as hydrophobic stacking interactions between the benzophenone, the aromatic ring of the boronic acid moiety, and the hydrophobic face of the carbohydrate. This interaction is stable enough such that the complex is able to be submitted to mass spectrometry without dissociating.

Fluorescent Glycosensors Monitored by Microarrays

To test the viability of the glycosensor across a large variety of carbohydrates, a fluorescent Alexa 555 dye was conjugated onto the terminal amine of the glycosensor, yielding compound **24**. For these experiments, cross-linking was not necessary, due to the omission of the benzophenone moiety and the results obtained from previous irradiation experiments (vide supra). Because the structure of Alexa 555 is proprietary to Life Technologies, Inc., the exact structure of **24** is not given.

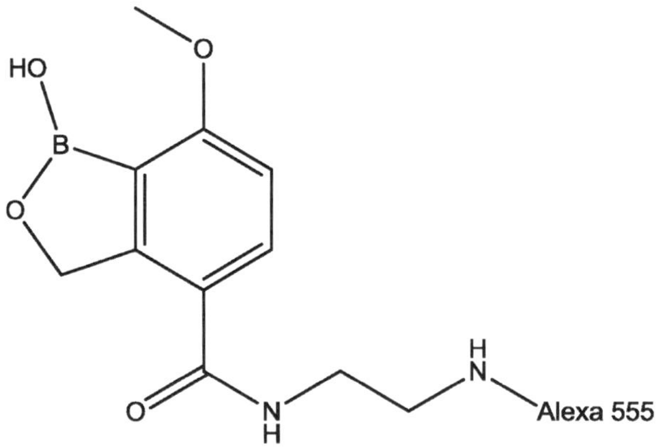

Figure A.19. Alexa 555 conjugated glycosensor **24**.

A glass slide with the printed glycan microarray was provided by Robotic Labware Designs (Encinitas, CA). Upon exposure of glycosensor **24** to the glycan array, strong binders can be visualized by their fluorescent tag, whereas weak binders are removed in subsequent washing steps. From the library of carbohydrates tested, the 17 strongest binders are reported (Figure A.20). There is no discernable pattern of carbohydrates that glycosensor **24** preferentially binds. The glycosensor appears to rely on the boronic acid moiety to bind to the carbohydrates non-specifically. The microarray data showed positive responses across most carbohydrates present on the slide, but only the compounds which yielded the strongest fluorescence are reported here.

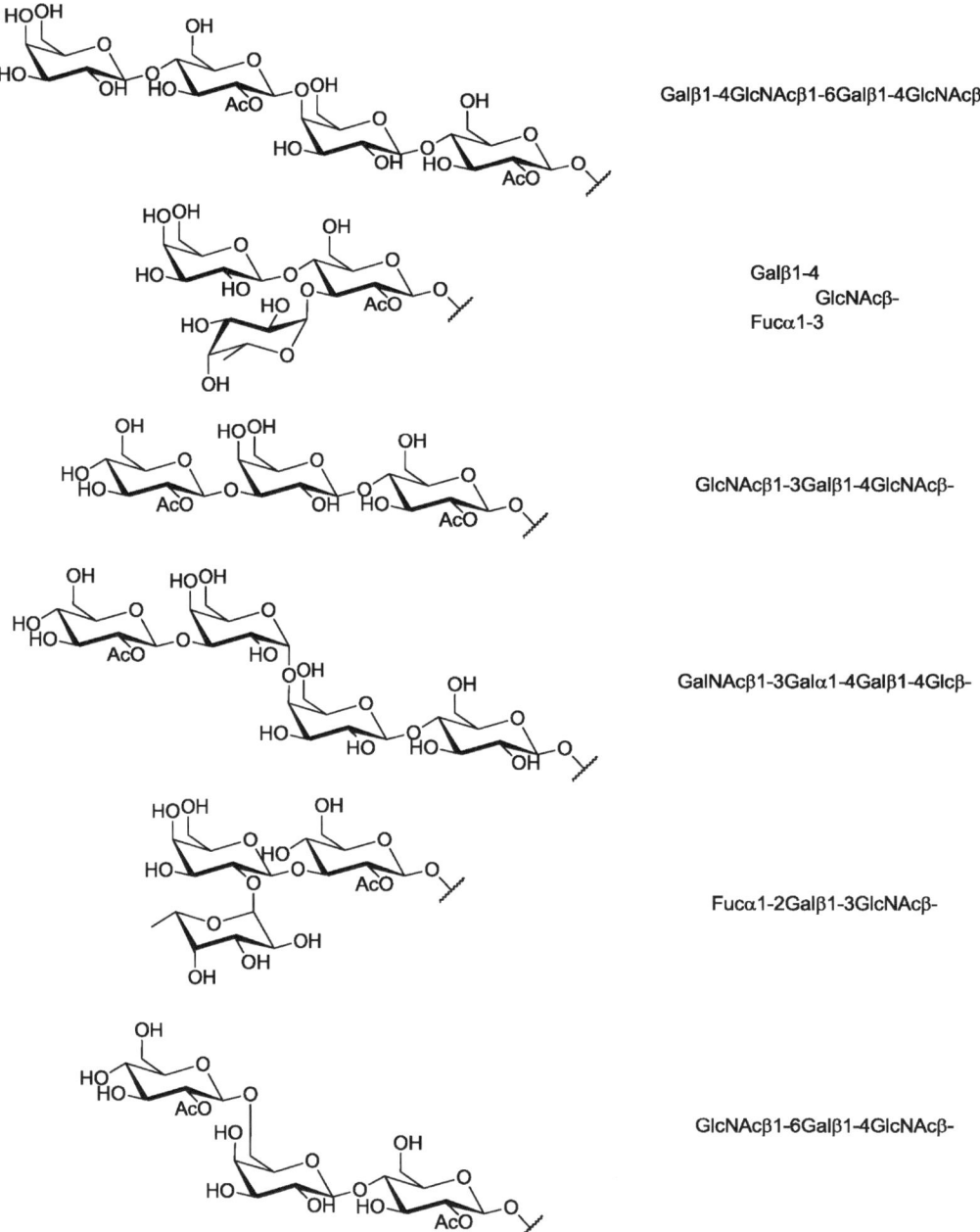

Figure A.20. Structures of microarray glycosides which consistently yielded strong fluorescence.

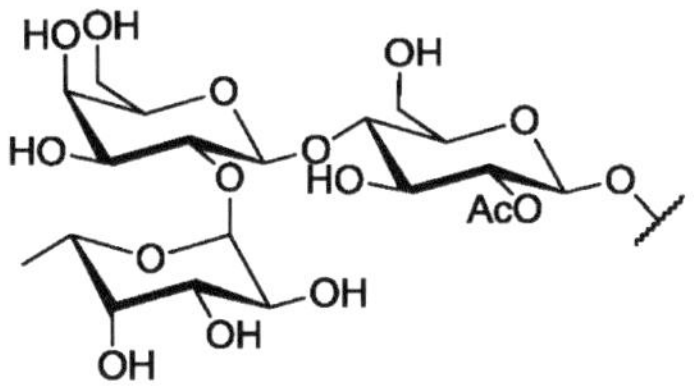

Fucα1-2Galβ1-4GlcNAcβ-

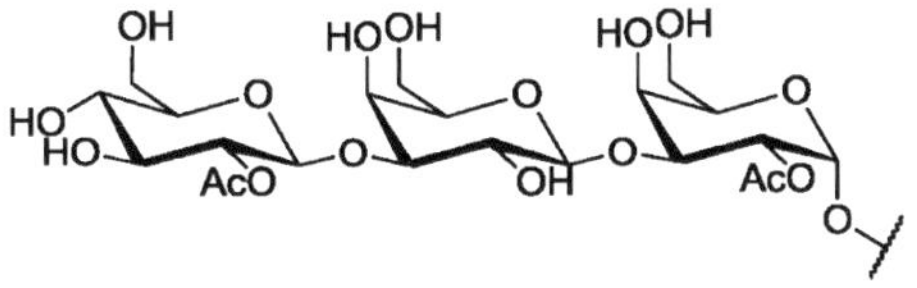

GlcNAcβ1-3Galβ1-3GalNAcα-

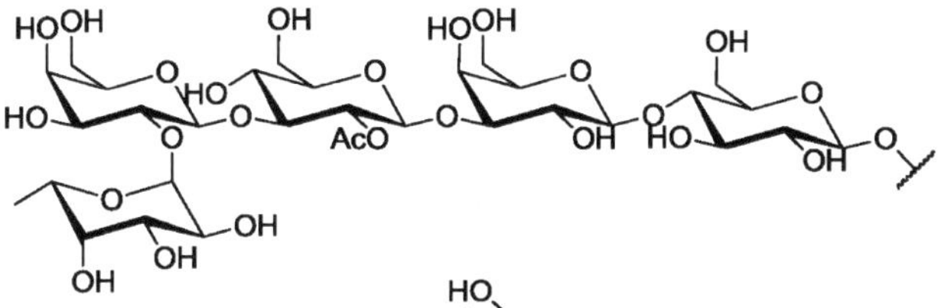

Fucα1-2Galβ1-3GlcNAcβ1-3Galβ1-4Glcβ-

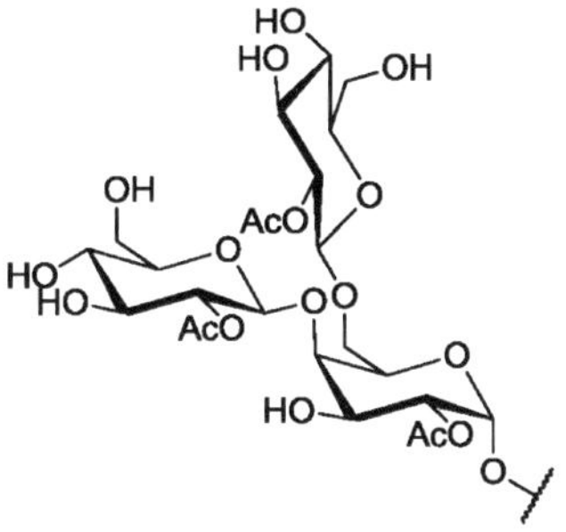

GlcNAcβ1-6
    GalNAcα-
GlcNAcβ1-4

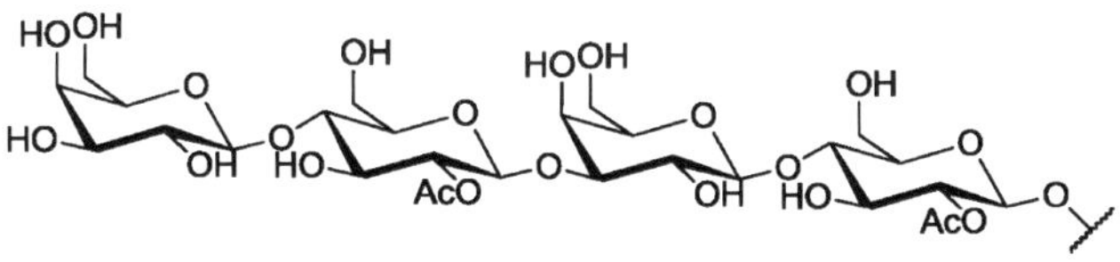

Galβ1-4GlcNAcβ1-3Galβ1-4GlcNAcβ-

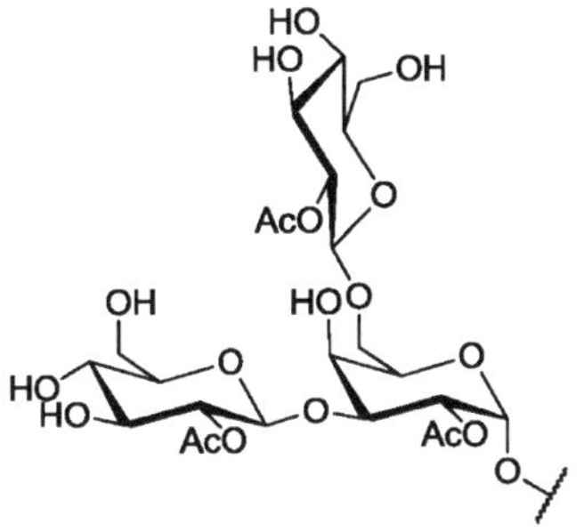

GlcNAcβ1-6
    GalNAcα-
GlcNAcβ1-3

Figure A.20. (continued)

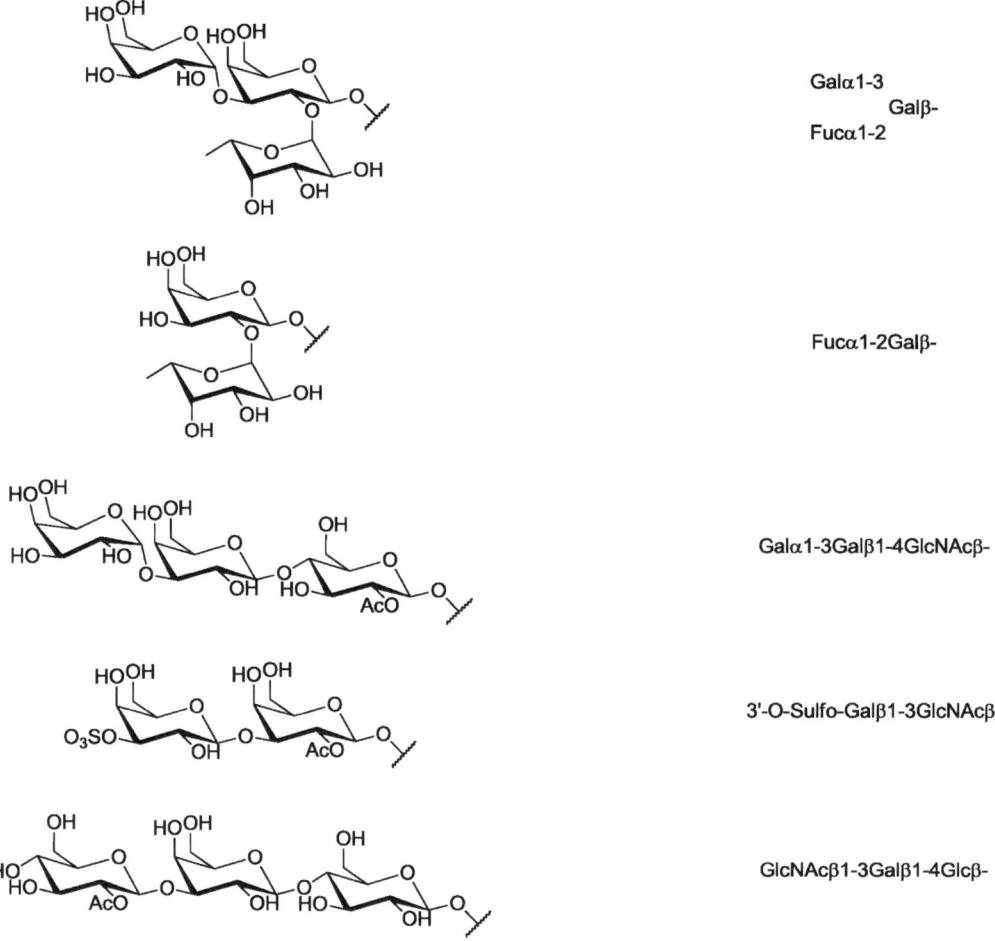

Figure A.20. (continued)

Conclusions

The boronic acid moiety has been demonstrated to bind to carbohydrates, but forming a covalent bond between a carbohydrate and the benzophenone moiety has been difficult. ARS and ^{1}H-NMR assays have shown that functionalization of the boronic acid structure renders the compound unsuitable for evaluation by these methods, as the ARS assay requires a shift in the optical properties upon binding and the ^{1}H-NMR assay requires a significant chemical shift upon binding. ITC was able to yield association constants, but requires a significant amount of material to elicit heats of binding and for the accurate determination of K_a. Mass Spectrometry of the benzophenone-coupled glycosensor was inconclusive to its ability to photoreactively couple to a carbohydrate. High-throughput screening of carbohydrates using a microarray confirmed that the glycosensor binds carbohydrates non-specifically. Efforts are currently ongoing to evaluate the binding and photoreactivity of the boronic acid/ benzophenone binding moiety with glycosides.

Experimentals

General Methods. General reagents were purchased from Acros. Fetuin was purchased from Sigma. Glycosensors were synthesized by the Grieco group and were used as described. Mass spectrometry analyses were submitted to Ohio State Mass Spectrometry & Proteomics Facility (Columbus, OH).

ARS Assays. A Quartz Fluorometer Cell (16.50F-Q-10/Z15) was purchased from Starna Cells, Inc. for use with the ARS Assays. Preliminary binding studies were performed on phenylboronic acid and *o*-hydroxymethyl phenylboronic acid according to a modified procedure as previously described.[200, 205] Briefly, 0.1 mM Alizarin Red S (ARS) was prepared in a 0.1 M phosphate solution buffered at pH 7.4 (Solution A). Boronic acid was added to a portion of solution A to make a 0.1 mM ARS, 10 mM boronic acid solution (Solution B). Solution B was diluted with solution A and absorbance values were determined at 450 nm.

The association constant for the boronic acid-ARS complex is the quotient of the intercept and the slope in a plot of 1/[boronic acid] vs. 1/ΔA.

Competitive ARS Assays. Competitive assays were performed similar to the above mentioned ARS assays. A solution of 0.1 mM ARS and 5 mM boronic acid was prepared in a 0.1 M phosphate solution buffered at pH 7.4 (Solution C). Fructose was added to a portion of solution C to make a 200 mM fructose solution (Solution D). Increments of solution D was added to solution C to make mixtures with a constant concentration of ARS and boronic acid and a range of concentrations of diol.

The K_{eq} is determined by plotting 1/P vs. Q, where

$$P = [L_0] - \frac{1}{QK_{ARS}} - \frac{[ARS_0]}{Q+1}$$

Equation A.1. Equation to determine K_a via the ARS Assay.

Q is a ratio of the concentration of free ARS to complexed ARS, determined by the change in absorbance of the solution; $[L_0]$ is the total amount of boronic acid; $[ARS_0]$ is the total amount of ARS; K_{ARS} is the association constant of the boronic acid-ARS complex. The K_a of the boronic acid-diol is then determined by dividing K_{ARS} by the slope of 1/P vs. Q.

ITC Assays. Binding constants and thermodynamic data were acquired using an isothermal titration calorimeter from Microcal, Inc. (Northhampton, MA). 40 Injections of 6 μL of 75 mM fructose or 75 mM sialic acid were added via a 250 μL syringe at an interval of 4 minutes into 1.435 mL of a 5 mM solution of the boronic acid while stirring at 310 rpm. Titrations were done at 25 °C in a 0.1 M phosphate buffered saline solution (pH 7.4).

Competitive ITC Assays. Competitive ITC experiments were performed similarly to a previously published procedure[206] using an isothermal titration calorimeter from Microcal, Inc. (Northhampton, MA). 40 Injections of 6 μL of 75 mM fructose were added via a 250 μL syringe at an interval of 4 minutes into 1.435 mL of a 5 mM solution of the boronic acid mixed with 50 mM O-methyl-N-acetylglucosamine while stirring at 310 rpm. Titrations were done at 25 °C in a 0.1 M phosphate buffered saline solution (pH 7.4).

Analysis of ITC data. Analysis by a nonlinear least squares curve fitting program (Origin 5.0) provided an apparent binding affinity K_{app}. For competitive ITC assays, K_a was determined using Equation A.2[206]:

$$K_2 = \left(\frac{K_1}{K_{app}} - 1\right) \times \frac{1}{L_{2tot}}$$

Equation A.2. Equation to determine K_a via competitive ITC.

Where K_2 is the binding affinity of the low-affinity ligand L_2 with a total concentration of L_{2tot}, K_1 is the binding affinity of the high affinity ligand titrated into the cell, and K_{app} is the apparent binding affinity obtained by titrating the high affinity ligand L_1 into the boronic acid in the presence of the low affinity ligand L_2.

NMR Assays. Binding constants were determined by ^{1}H-NMR titration experiments as described by Dowlut and Hall.[200] A deuterated buffer solution was made by evaporating an aqueous solution of 10 mM phosphate buffered saline, pH 7.4, and adding an equal volume of D_2O to yield a deuterated buffer of the same concentration, adjusting the pH as necessary. The boronic acid (either *o*-hydroxymethyl phenylboronic acid or **2**) was dissolved in the deuterated buffer to yield a 15 mM boronic acid solution (solution A). A 150 mM solution of fructose was made by dissolving 150 mM of sugar in 1 mL of solution A, adjusting the pH to 7.4 as necessary (solution B). This ensured the concentration of the boronic acid was constant with additions of sugar.

To 700 μL of solution A was added 40 μL of solution B, and afterwards 10 μL aliquots were added to obtain sugar concentrations in the range of 8-20 mM until the boronic acid solution was saturated with sugar. The complexed boronic acid/free acid ratio was measured directly from expanded integrals of relevant peaks on the ^{1}H-NMR spectra. A plot of 1/θ versus 1/[S] (Equation A.4) would give the K_a value as the inverse of the intercept.

Assuming a one-site binding for the equilibrium between a boronic acid receptor (R) and a saccharide substrate (S),

$$R + S \rightleftharpoons RS$$

The ratio [RS]/[R] was found from the ratio of peaks θ in the ^{1}H-NMR spectrum.

$$\theta = \frac{[RS]}{[RS] + [R]}$$

Equation A.3. Ratio of free and complexed boronic acid peaks determined by ^{1}H-NMR Assay.

The association constant K_a was determined from the Benesi-Hildebrand plot:

$$1/\theta = 1/K_a(1/[S]) + 1$$

Equation A.4. Equation to determine K_a via ^{1}H-NMR Assay.

where $1/\theta = 1 + [RS]/[R]$ and [S] is the substrate concentration.

Cross-linking experiments with benzophenone-containing glycosensors. For experiments using glycoprotein, 45 μL of a 50 μM solution of the glycosensor were mixed with 5 μL of a 2 mg/mL solution of fetuin in H$_2$O and added to a 2 mL microcentrifuge tube placed in an ice bath. The solution was placed 5-10 cm from a 15 Watt lamp and irradiated for 30 min at 365 nm. This was done with the cap of the microcentrifuge tube open, to prevent filtering effects. As a control, the experiment was repeated without irradiation.

For experiments using carbohydrates, 50 μL of a 50 μM solution of the glycosensor and 50 μL of 50 μM carbohydrate containing solution were added to a 2 mL microcentrifuge tube and placed in an ice bath. The solution was placed 5-10 cm from a 15 Watt lamp and irradiated for 30 min at 365 nm. This was done with the cap of the microcentrifuge tube open, to prevent filtering effects. As a control, the experiment was repeated without irradiation.

Microarray Assays. A glycan-printed microarray was provided by Robotic Labware Designs, Inc. (Encinitas, CA). The location of the glycan area of the microarray was noted as a tracing reference, as it became difficult to see after the next step. The remaining NHS groups present on the slide were blocked with 50 mM ethanolamine in 50 mM borate buffer, pH 9.2 for one hour, using a slide rack to fully submerge the slides. The slides were then rinsed well by repeated submersion in three separate containers filled with deionized water. The slides were quickly dried by centrifugation at 1200 rpm for one minute, and then set to dry additionally in a desiccator for several hours. Once sufficiently dry, the outside borders of the glycan area were marked with a PAP pen, based on the prepared trace reference. The slides were then stored in the desiccator at room temperature for at least 8 hours and a maximum of 7 days before use to minimize background fluorescence.

Before use, the slides were rehydrated by submersion in 10 mM PBS, pH 7.4 containing 1% tween-20 for one hour. The slide was dried by blotting the edges on a paper towel followed immediately by additional drying through centrifugation at 1200 rpm for one minute. The dye conjugate was dissolved to approximately 50 μg/mL with 10 mM PBS, pH 7.4 containing 0.1% tween-20. This solution was incubated on the slide surface at room temperature for 1 hour with orbital motion, ensuring the solution remains hydrated on the slide

The slides were then well rinsed by repeated submersion in a container filled with 10 mM PBS, pH 7.4 with 0.1% tween-20, followed by repeated submersion in 10 mM PBS, pH 7.4 with 0.001% tween-20, followed by repeated submersion in a separate container filled with deionized water. The slides were then immediately dried by centrifugation at 1200 rpm for one minute and submitted to Robotic Labware Designs, Inc. for fluorescence analysis.

The strongest binders were determined by marking the carbohydrate binding partners with fluorescence above an arbitrary intensity (2000). Carbohydrates which showed an intensity above the cutoff for at least 10 of the 16 arrays were considered 'hits' and labeled as a strong binder.

Microarray Assays with crosslinking. A glycan-printed microarray was provided by Robotic Labware Designs, Inc. (Encinitas, CA). The location of the glycan area of the microarray was noted as a tracing reference, as it became difficult to see after the next step. The remaining NHS groups present on the slide were blocked with 50 mM ethanolamine in 50 mM borate buffer, pH 9.2 for one hour, using a slide rack to fully submerge the slides. The slides were then rinsed well by repeated submersion in three separate containers filled with deionized water. The slides were quickly dried under a stream of Argon, and then set to dry additionally in a desiccator for several hours. Once sufficiently dry, the outside borders of the glycan area were marked with a PAP pen, based on the prepared trace reference. The slides were then stored in the desiccator at room temperature for at least 8 hours and a maximum of 7 days before use to minimize background fluorescence.

Before use, the slides were rehydrated by submersion in 10 mM PBS, pH 7.4 containing 0.001% tween-20 for one hour. The slide was dried by blotting the edges on a paper towel followed immediately by additional drying under an Argon stream. The dye conjugate was dissolved to approximately 50 μg/mL with 10 mM PBS, pH 7.4 containing 0.001% tween-20. This solution was incubated on the slide surface at room temperature for 2 hours, ensuring the solution remains hydrated on the slide. To cross-link the dye conjugate onto the slide surface, the slide was carefully placed onto a dish swimming in an ice bath, and irradiated at 365 nm for 30 minutes at 0°C. The control slide was incubated without irradiation for 30 minutes.

The slides were then well rinsed by repeated submersion in a container filled with 10 mM PBS, pH 7.4 with 0.1% tween-20, followed by repeated submersion in 10 mM PBS, pH 7.4 with 0.001% tween-20, followed by repeated submersion in a separate container filled with 10 mM PBS, pH 7.4 with 0.001% tween-20. The slides were then immediately dried under an Argon stream and submitted to Robotic Labware Designs, Inc. for fluorescence analysis. Analysis of these slides was unsuccessful, so repetition of this experiment at Robotic Labware Designs, Inc. is currently ongoing.

REFERENCES CITED

1. Apweiler, R.; Hermajakob, H.; Sharon, N., On the Frequency of Protein Glycosylation, As Deducted from Analysis of the SWISS-PROT Database. *Biochim. Biophys. Acta* **1999**, *1473*, 4-8.

2. Wang, C.; Eufemi, M.; Turano, C.; Giartorsio, A., Influence of the Carbohydrate Moiety on the Stability of Glycoproteins. *Biochemistry* **1996**, *35*, 7299-7307.

3. Dwek, R. A., Glycobiology: Toward Understanding the Function of Sugars. *Chem. Rev.* **1996**, *96*, 683-720.

4. Landsteiner, K., Zur Kenntnis der Antifermentativen, Lytischen, und Agglutinierenden Wirkungen des Blutserum und der Lymphe. *Zbl. Bakt.* **1900**, *27*, 357-362.

5. Greenwell, P., Blood Group Antigens: Molecules Seeking a Function? *Glycoconjugate J.* **1997**, *14*, 159-173.

6. Shriver, Z.; Raguram, S.; Sasisekharan, R., Glycomics: A Pathway to a Class of New and Improved Therapeutics. *Nat. Rev. Drug Discov.* **2004**, *3*, 863-873.

7. Aoki, K. F.; Yamaguchi, A.; Okuno, Y.; Akutsu, T.; Ueda, N.; Kanehisa, M.; Mamitsuka, H., Efficient Tree-Matching Methods for Accurate Carbohydrate Database Queries. *Genome Inform.* **2003**, *14*, 134-143.

8. Comelli, E. M.; Head, S. R.; Gilmartin, T.; Whisenant, T.; Haslam, S. M.; North, S. J.; Wong, N. K.; Kudo, T.; Narimatsu, H.; Esko, J. D.; Drickamer, K.; Dell, A.; Paulson, J. C., A Focused Microarray Approach to Functional Glycomics: Transcriptional Regulation of the Glycome. *Glycobiology* **2006**, *16*, 117-131.

9. Hsu, K. L.; Pilobello, K. T.; Mahal, L. K., Analyzing the Dynamic Bacterial Glycome with a Lectin Microarray Approach. *Nat. Chem. Biol.* **2006**, *2*, 153-157.

10. Vocadlo, D. J.; Hang, H. C.; Kim, E. J.; Hanoever, J. A.; Bertozzi, C. R., A Chemical Approach for Identifying *O*-GlcNAc Modified Proteins in Cells. *Proc. Natl. Acad. Sci. USA.* **2003**, *100*, 9116-9121.

11. An, H. J.; Miyamoto, S.; Lancaster, K. S.; Kirmiz, C.; Li, B.; Lam, K. S.; Leiserowitz, G. S.; Lebrilla, C. B., Profiling of Glycans in Serum for the Discovery of Potential Biomarkers for Ovarian Cancer. *J. Proteome Res.* **2006**, *5*, 1626-1635.

12. Gorelik, E.; Galili, U.; Raz, A., On the Role of Cell Surface Carbohydrates and Their Binding Proteins (Lectins) in Tumor Metastasis. *Cancer Metastasis Rev.* **2001**, *20*, 245-277.

13. Rudd, P. M.; Wormald, M. R.; Dwek, R. A., Sugar-Mediated Ligand-Receptor Interactions in the Immune System. *Trends Biotechnol.* **2004**, *22*, 524-530.

14. Fischer, E., Einfluss der Konfiguration auf die Wirkung der Enzyme. *Ber. Dtsch. Chem. Ges.* **1894**, *27*, 2984-2993.

15. Kunz, H., Emil Fischer – Unequalled Classicist, Master of Organic Chemistry Research, and Inspired Trailblazer of Biological Chemistry. *Angw. Chem. Int. Edit.* **2002**, *41*, 4439-4451.

16. Blake, C. C.; Johnson, L. N.; Mair, G. A.; North, A. C.; Phillips, D. C.; Sarma, R., Crystallographic Studies of the Activity of Hen Egg-White Lysozyme. *Proc. R. Soc. B.* **1967**, *167*, 378-388.

17. Fleming, A., On a Remarkable Bacteriolytic Element Found in Tissues and Secretions. *Proc. R. Soc. B.* **1922**, *93*, 306-317.

18. Hardman, K. D.; Ainsworth, C. F., Structure of Concanavalin A at 2.4 Å Resolution. *Biochemistry* **1972**, *11*, 4910-4919.

19. Wilson, K. A.; Skehel, J. J.; Wiley, D. C., Structure of the Haemagglutinin Membrane Glycoprotein of the Influenza Virus at a 3 Å Resolution. *Nature* **1981**, *289*, 366-373.

20. Varki, A.; Cummings, R. D.; Esko, J. D.; Freeze, H. H.; Stanley, P.; Bertozzi, C. R.; Hart, G. W.; Etzler, M. E., *Essentials of Glycobiology, 2nd Ed.* Cold Spring Harbor Laboratory Press: Cold Spring Harbor, N. Y., 2009.

21. Rao, V. S. R.; Qasba, P. K.; Balaji, P. V.; Chandrasekaran, R., *Conformation of Carbohydrates* Harwood Academic Publishers: The Netherlands, 1998.

22. Merrit, E. A.; Kuhn, P.; Sarfaty, S.; Erbe, J. L.; Holmes, R. K.; Hol, W. G., The 1.25 Å Resolution Refinement of the Cholera Toxin B Pentamer: Evidence of Peptide Backbone Strain at the Receptor-Binding Site. *J. Mol. Biol.* **1998**, *282*, 1043-1059.

23. Merrit, E. A.; Sarfaty, S.; van der Akker, F.; L'Hoir, C.; Martial, J. A.; Hol, W. G., Crystal Structure of Cholera Toxin B-Pentamer Bound to Receptor GM1 Pentasaccharide. *Protein Sci.* **1994**, *3*, 166-175.

24. Pusztai, A., *Plant Lectins*. Cambridge University Press: Cambridge, 1991.

25. Laine, R. A., *Glycosciences: Status and Perspectives*. Chapman and Hall: London, 1997.

26. Gabius, H.; Siebert, H.; André, H.; Jiménez-Barbero, J.; Rüdiger, H., Chemical Biology of the Sugar Code. *J. Mol. Biol.* **2004**, *5*, 740-764.

27. Mitchell, S. W., Researches Upon the Venom of the Rattlesnake. *Smithsonian Contrib. Knowledge* **1860**, *12*, 89-90.

28. Stillmark, H., Über Ricin, ein Giftiges Ferment aus den Samen von *Ricinus Comm. L.* und Einigen Anderen Euphorbiaceen. Inaugural Dissertation, Schnakenburg's Buchdruckerei, Dorpat, 1888.

29. Sumner, J. B., The Globulins of the Jack Bean, Canavalia Ensiformis: Preliminary Paper. *J. Biol. Chem.* **1919**, *37*, 137-142.

30. Boyd, W. C., *The Proteins.* Academic Press: New York, 1954.

31. Bittiger, H.; Schnebli, H. P., *Concanavalin A as a Tool.* John Wiley & Sons: New York, 1976.

32. Naismith, J. H.; Emmerich, C.; Habash, J.; Harrop, S. J.; Helliwell, J. R.; Hunter, W. N.; Raferty, J.; Kalb(Gilboa), A. J.; Yariv, J., Refined Structure of Concanavalin-A Complexed with α-Methyl-D-Mannopyranoside at 2.0 Angstrom Resolution and Comparison with the Saccharide-Free Structure. *Acta Crystallogr. D* **1994**, *50*, 847-858.

33. Drickamer, K., Two Distinct Classes of Carbohydrate-Recognition Domains in Animal Lectins. *J. Biol. Chem.* **1988**, *263*, 9557-9560.

34. Zapp, H.; Snell, M. E.; Bossard, M. J., PEGylation of Cyanovirin-N, an Inhibitor of HIV. *Adv. Drug. Rev.* **2008**, *60*, 79-87.

35. van den Brule, F. A.; Liu, F. T.; Castronovo, V., Transglutaminase-Mediated Oligomerization of Galectin-3 Modulates Human Melanoma Cell Interactions with Laminin, *Cell Adhes. Commun.* **1998**, *5*, 425-435.

36. Henderson, N. C.; Mackinnon, A. C.; Farnworth, S. L.; Poirier, F.; Russo, F. P.; Iredale, J. P.; Haslett, C.; Simpson, K. J.; Sethi, T., Galectin-3 Regulates Myofibroblast Activation and Hepatic Fibrosis. *Proc. Natl. Acad. Sci. USA.* **2006**, *103*, 5060-5065.

37. Kasper, M.; Hughes, R. C., Immunocytochemical Evidence for a Modulation of Galectin-3 (Mac-2), a Carbohydrate Binding Protein, in Pulmonary Fibrosis. **1996**, *179*, 309-316.

38. Hsu, D. K.; Dowling, C. A.; Jeng, K. C. G.; Chen, J. T.; Yang, R. Y.; Liu, F. T., Galectin-3 Expression is Induced in Cirrhotic Liver and Hepatocellular Carcinoma. *International Journal of Cancer* **1999**, *81*, 519-526.

39. Lotan, R.; Ito, H.; Yasui, W.; Yokozaki, H.; Lotan, D.; Tahara, E., Expression of a 31-kDa Lactoside-Binding Lectin in Normal Human Gastric-Mucosa and in Primary and Metastatic Gastric Carcinomas. *International Journal of Cancer* **1994**, *56*, 474-480.

40. Lee, Y. C.; Lee, T. L., Carbohydrate-Protein Interactions: Basis of Glycobiology. *Acc. Chem. Res.* **1995**, *28*, 321-327.

41. Mammen, M.; Choi, S.-K.; Whitesides, G. M., Polyvalent interactions in biological systems: Implications for design and use of multivalent ligands and inhibitors *Angew. Chem. Int. Ed.* **1998**, *37*, 2754–2794.

42. Lee, R. T.; Lee, Y. C., Affinity enhancement by multivalent lectin-carbohydrate interaction" *Glycoconjugate J.* **2000**, *17*, 543–551.

43. Kilpatrick, D. C., *Handbook of Animal Lectins: Properties and Biomedical Applications* Wiley-VCH: New York, 2000.

44. Lundquist, J. J.; Toone, E. J., The Cluster Glycoside Effect. *Chem. Rev.* **2002**, *102*, 555-578.

45. Wolfenden, M. L., *Using PAMAM Dendrimer Frameworks to Investigate Multivalent Binding in Protein : Carbohydrate Interactions.* **2009**, Ph. D. Thesis, Montana State University.

46. Gestwicki, J. E.; Cairo, C. W.; Strong, L. E.; Oekiten, K. A.; Kiessling, L. L., Influencing Receptor-Ligand binding Mechanisms with Multivalent Ligand Architecture. *J. Am. Chem. Soc.* **2002**, *124*, 14922-14933.

47. Jevprasesphant, R.; Penny, J.; Jalal, R.; Attwood, D.; McKeown, N. B.; D'Emanuele, A., The Influence of Surface Modification on the Cytotoxicity of PAMAM Dendrimers. *Int. J. Pharm.* **2003**, *252*, 263-266.

48. Glick, G. D.; Knowles, J. R., Molecular Recognition of Bivalent Sialosides by Influenza Virus. *J. Am. Chem. Soc.* **1991**, *113*, 4701-4703.

49. Pagé, D.; Roy, R., Optimizing Lectin-Carbohydrate Interactions: Improved Binding of Divalent α-mannosylated Ligands toward Concanavalin A. *Glycoconjugate J.* **1997**, *14*, 345-356.

50. Lindhorst, T. K.; Kieberg, C.; Krallmann-Wenzel, U., Inhibition of the Type 1 Fimbriae-Mediated Adhesion of Escherichia Coli to Erythrocytes by Multiantennary α-mannosyl clusters: The Effect of Multivalency. *Glycoconjugate J.* **1998**, *15*, 605-613.

51. Kanai, M.; Mortell, K. H.; Kiessling, L. L., Varying the Size of Multivalent Ligands: The Dependence of Concanavalin A Binding on Neoglycopolymer Length. *J. Am. Chem. Soc.* **1997**, *119*, 9931-9932.

52. Woller, E. K.; Walter, E. D.; Morgan, J. R.; Singel, D. J.; Cloninger, M. J., Altering the Strength of Lectin Binding Interactions and Controlling the Amount of Lectin Clustering Using Mannose/Hydroxyl-Functionalized Dendrimers. *J. Am. Chem. Soc.* **2003**, *125*, 8820-2226.

53. Wolfenden, M. L.; Cloninger, M. J., Mannose/Glucose Dendrimers to Investigate the Predictable Tunability of Multivalent Interactions. *J. Am. Chem. Soc.* **2005**, *127*, 12168-12169.

54. Choi, S-K., *Synthetic Multivalent Molecules* Wiley-VCH: US, 2004.

55. Newkome, G. R.; Moorefield, C. N.; Vögtle, F., *Dendrimers and Dendrons: Concepts, Syntheses, Applications* Wiley-VCH: Weinheim, 2001.

56. Buhleier, E., Wehner, W.; Vögtle, F., 'Cascade' and 'Nonskid-Chain-like' Syntheses of Molecular Cavity Topologies. *Synthesis* **1978**, 155-158.

57. Dykes, G. M., Dendrimers: A Review of Their Appeal and Applications. *J. Chem. Technol. Biotechnol.* **2001**, *76*, 903-918.

58. Tomalia, D. A.; Naylor, A. M.; Goddard, W. A., Starburst Dendrimers - Molecular-level Control of Size Shape, Surface Chemistry, Topology, and Flexibility from Atoms to Macroscopic Matter. *Angew. Chem. Int. Ed.* **1990**, *29*, 138-175.

59. Fischer, M.; Vögtle, F., Dendrimers: From Design to Application – A Progress Report. *Angew. Chem. Int. Ed.* **1999**, *38*, 885-905.

60. Zeng, F.; Zimmermann, S. C., Dendrimers in Supramolecular Chemistry: From Molecular Recognition to Self-Assembly. *Chem. Rev.* **1997**, *97*, 1681-1712.

61. Morgan, J. R., *The Synthesis of Glycodendrimers and Their Applications in Carbohydrate-Protein Interactions and Catalysis.* **2006**, Ph. D. Thesis, Montana State University.

62. Newkome, G. R.; Gupta, V. K.; Baker, G. R.; Yao, Z.-Q., Micelles. Part 1. Cascade Molecules: A New Approach to Micelle. A [27]-Arborol. *J. Org. Chem.* **1985**, *50*, 2003.

63. Tomalia, D. A., Starburst/Cascade Dendrimers: Fundamental Building Blocks for a New Nanoscopic Chemistry Set. *Aldrichimica Acta* **1993**, *26*, 91-101.

64. Hawker, C. J.; Fréchet, J. M. J., Preparation of Polymers with Controlled Dendritic Architecture. A New Convergent Approach to Dendritic Macromolecules. *J. Am. Chem. Soc.* **1990**, *112*, 7638-7647.

65. James, T. D.; Shinmori, H.; Takeuchi, M.; Shinkai, S., A Saccharide 'Sponge'. Synthesis and Properties of a Dendritic Boronic Acid. *Chem. Commun.* **1996**, 705-706.

66. Newkome, G. R.; Moorefield, C. N.; Baker, G. R.; Saunders, M. J.; Grossman, S. H., Alkane Cascade Polymers Possessing Micellar Topology: Micellenaoic Acid Derivatives. *Angew. Chem. Int. Ed.* **1999**, *30*, 1178-1180.

67. Hawker, C. J.; Wooley, K. L.; Fréchet, J. M. J., Unimolecular Micelles and Globular Amphiphiles – Dendritic Macromolecules as Novel Solubilization Agents. *J. Chem. Soc., Perkin Trans. 1* **1993**, *21*, 1287-1297.

68. Devadoss, C.; Bharathi, P.; Moore, J. S., Energy Transfer in Dendritic Macromolecules: Molecular Size Effects and the Role of an Energy Gradient. *J. Am. Chem. Soc.* **1996**, *118*, 9635-9644.

69. Jiang, D. L.; Aida, T., Photoisomerization in Dendrimers by Harvesting of Low-Energy Photons. *Nature* **1997**, *388*, 454-456.

70. Boas, U.; Heegaard, P. M. H., Dendrimers in Drug Research *Chem. Soc. Rev.* **2004**, *33*, 43–63.

71. Klemm, J. D.; Schreiber, S. L.; Crabtree, G. R., Dimerization as a Regulatory Mechanism in Signal Transduction. *Ann. Rev. Immunol.* **1998**, *16*, 569-592.

72. Pace, K. E.; Lee, C.; Stewart, P. L.; Baum, L. G., Restricted Receptor Segregation into Membrane Microdomains Occurs on Human T Cells During Apoptosis Induced by Galectin-1. *J. Immunol.* **1999**, *163*, 3801-3811.

73. Uversky, V. N.; Fink, A., *Protein Misfolding, Aggregation and Conformational Diseases: Part A: Protein Aggregation and Conformational Diseases*. Springer Verlag: New York, 2006.

74. Fersht, A. R., *Structure and Mechanism in Protein Science. A Guide to Enzyme Catalysis and Protein Folding.* W. H. Freeman & Company: New York, 1999.

75. Garcia-Mata, R.; Gao, Y.-S., Sztul, E., Hassles with Taking Out the Garbage: Aggrevating Aggresomes. *Traffic* **2002**, *3*, 388-396.

76. Wickner, S.; Maurizi, M. R.; Gottesman, S., Posttranslational Quality Control: Folding, Refolding, and Degrading Proteins. *Science* **1999**, *286*, 1888-1893.

77. Ron, D.; Walter, P., Signal Integration in the Endoplasmic Reticulum Unfolded Protein Response. *Nat. Rev. Mol. Cell Biol.* **2007**, *8*, 519-529.

78. Rodriguez-Gonzalez, A.; Lin, T.; Ikeda, A. K.; Simms-Waldrip, T.; Fu, C.; Sakamoto, K. M., Role of the Aggresome Pathway in Cancer: Targeting Histone Deacetylase 6-Dependent Protein Degradation. *Cancer Res.* **2008**, *68*, 2557-2560.

79. Meriin, A. B.; Gabai, V. L.; Yaglom, J.; Shifrin, V. I.; Sherman, M. Y., Proteasome Inhibitors Stress Kinases and Induce Hsp72. Diverse Effects on Apoptosis. *J. Biol. Chem.* **1998**, *273*, 6373-6379.

80. Johnston, J. A.; Ward, C. L.; Kopito, R. R., Aggresomes: A Cellular Response to Misfolded Proteins. *J. Cell. Biol.* **1998**, *143*, 1883-1898.

81. Brujin, L. I.; Houseweart, M. K.; Kato, S.; Anderson, K. L.; Anderson, S. D.; Ohama, E.; Reaume, A. G.; Scott, R. W.; Cleveland, D. W., Aggregation and Motor Neuron Toxicity of an ALS-Linked SOD1 Mutant Independent from Wild-Type SOD1. *Science* **1998**, *281*, 1851-1854.

82. Masliah, E.; Rockenstein, E.; Veinbergs, I.; Mallory, M.; Hashimoto, M.; Takeda, A.; Sagara, Y.; Sisk, A.; Mucke, L., Dopaminergic Loss and Inclusion Body Formation in Alpha-Synuclein Mice: Implications for Neurodegenerative Disorders. *Science* **2000**, *287*, 1265-1269.

83. Davies, S. W.; Turmaine, M.; Cozens, B. A.; DiFiglia, M.; Sharp, A. H.; Ross, C. A.; Scherzinger, E.; Wanker, E. E.; Mangiarini, L.; Bates, G. P., Formation of Neuronal Intranuclear Inclusions Underlies the Neurological Dysfunction in Mice Transgenic for the HD Mutation. *Cell* **1997**, *90*, 537-548.

84. Lin, C. H.; Tallaksen-Greene, S.; Chien, W. M.; Cearly, J. A.; Jackson, W. S.; Crouse, A. B.; Ren, S.; Li, X. J.; Albin, R. J.;Detloff, P. J., Neurological Abnormalities in a Knock-In Mouse Model of Huntington's Disease. *Hum. Mol. Genet.* **2001**, *10*, 137-144.

85. Zoghbi, H. Y.; Orr, H. T., Glutamine Repeats and Neurodegeneration. *Annu. Rev. Neurosci.* **2000**, *23*, 217-247.

86. Johnston, J. A.; Dalton, M. J.; Gurney, M. E.; Kopito, R. R., Formation of High Molecular Weight Complexes of Mutant Cu, Zn-Superoxide Dismutase in a Mouse Model for Familial Amyotrophic Lateral Sclerosis. *Proc. Natl. Acad. Sci. USA* **2000**, *97*, 12571-12576.

87. Rajan, R. S.; Illing, M. E.; Bence, N. F.; Kopito, R. R., Specificity in Intracellular Protein Aggregation and Inclusion Body Formation. *Proc. Natl. Acad. Sci. USA* **2001**, *98*, 13060-13065.

88. Zhang, B.; Tu, P.; Abtahian, F.; Trojanowski, J. Q.; Lee, V. M., Neurofilaments and Orthograde Transport Are Reduced in Ventral Root Axons of Transgenic Mice That Express Human SOD1 with a G93A Mutation. *J. Cell. Biol.* **1997**, *139*, 1307-1315.

89. Stranger, B. E.; Forrest, M. S.; Dunning, M.; Ingle, C. E.; Beazley, C.; Thorne, N.; Redon, R.; Bird, C. P.; de Grassi, A.; Lee, C.; Tyler-Smith, C.; Carter, N.; Scherer, S. W.; Tavaré, S.; Deloukas, P.; Hurles, M. E.; Dermitzakis, E. T., Relative Impact of Nucleotide and Copy Number Variation on Gene Expression Phenotypes. *Science* **2007**, *315*, 848-853.

90. Sopko, R.; Huang, D.; Preston, N.; Chua, G.; Papp, B.; Kafadar, K.; Snyder, M.; Oliver, S. G.; Cyert, M.; Hughes, T. R.; Boone, C.; Andrews, B., Mapping Pathways and Phenotypes by Systematic Gene Overexpression. *Mol. Cell* **2006**, *21*, 319-330.

91. Vavouri, T; Semple, J. I.; Garcia-Verdugo, R.; Lehner, B., Intrinsic Protein Disorder and Interaction Promiscuity Are Widely Associated with Dosage Sensitivity. *Cell* **2009**, *138*, 198-208.

92. DeKosky, S. T.; Marek, K., Looking Backward to Move Forward: Early Detection of Neurodegenerative Disorders. *Science* **2003**, *302*, 830-834.

93. Nesterov, E. E.; Skoch, J.; Hyman, B. T.; Klunk, W. E.; Bacskai, B. J.; Swager, T. M., In Vivo Optical Imaging of Amyloid Aggregates in Brain: Design of Fluorescent Markers. *Angew. Chem. Int. Ed.* **2005**, *44*, 5452-5456.

94. Montine, T. J.; Quinn, J. F.; Milatovic, D.; Silbert, L. C.; Dang, T.; Sanchez, S.; Terry, E.; Roberts II, L. J.; Kaye, J. A.; Morrow, J. D., Peripheral F2-Isoprostanes Are Not Increased in Alzheimer's Disease. *Ann. Neurol.* **2002**, *52*, 175-179.

95. Frank, R. A.; Galasko, D.; Hampel, H.; Hardy, J.; de Leon, M. J.; Mehta, P. D.; Rogers, J.; Siemers, E.; Trojanowski, J. Q., Biological Markers for Therapeutic Trials in

Alzheimer's Disease: Proceedings of the Biological Markers Working Group; NIA Initiative on Neuroimaging in Alzheimer's Disease. *Neurobiol. Aging* **2003**, *24*, 521-536.

96. Nilsson, C. L., Lectins: Proteins that interpret the sugar code. *Anal. Chem.* **2003**, 349A–353A.

97. Loris, R.; Hamelryck, T.; Bouckaert, J.; Wyns, L., Legume lectin structure. *Biochim. Biophys. Acta* **1998**, *1383*, 9–36.

98. van Damme, E. J. M.; Peumans, W. J.; Barre, A.; Rouge, P., Plant lectins: A composite of several distinct families of structurally and evolutionary related proteins with diverse biological roles. *Crit. Rev. Plant Sci.* **1998**, *17*, 575–692.

99. Derewenda, Z.; Yariv, J.; Helliwell, J. R.; Kalb, A. J.; Dodson, E. J.; Papiz, M. Z.; Wan, T.; Campbell, J., The structure of the saccharide-binding site of Concanavalin A. *EMBO J.* **1989**, *8*, 2189–2193.

100. Trowbridge, I. S. Isolation and Chemical Characterization of a Mitogenic Lectin from *Pisum sativum.*, *J. Biol. Chem.* **1974**, *249*, 6004–6012.

101. Ruzheinikov, S. N.; Mikhailova, I. Y.; Tsygannik, I. N.; Pangborn, W.; Duax, W.; Pletnev, V. Z., The Structure of the Pea Lectin–D-mannopyranose Complex at a 2.1 Å Resolution. *J. Bioorg. Chem.* **1998**, *24*, 277–279.

102. Schwarz, F. P.; Puri, K. D.; Bhat, R. G.; Surolia, A. Thermodynamics of Monosaccharide Binding to Concanavalin A, Pea (*Pisum sativum*) Lectin, and Lentil (*Lens culinaris*) Lectin., *J. Biol. Chem.* **1993**, *268*, 7668–7677.

103. Van Eijsden, R. R.; De Pater, B. S.; Kijne, J. W., Mutational Analysis of the Sugar-binding site of Pea Lectin. *Glycoconjugate J.* **1994**, *11*, 375-380.

104. Roy, R., A decade of glycodendrimer chemistry. *Trends Glycosci. Glycotech.* **2003**, *15*, 291–310.

105. Woller, E. K.; Cloninger, M. J., The Lectin-Binding Properties of Six Generations of Mannose-Functionalized Dendrimers. *Org. Lett.* **2002**, *4*, 7–10.

106. Binding motifs are described in detail in reference 105. We define multivalent binding interactions as at least two sugars per dendrimer binding into two receptor sites on a lectin. Statistical or proximity enhancements upon affinity are smaller binding effects caused by the increased concentration of sugar near the binding site, are comparable to Effective Molarity phenomena of synthetic reactions, and are also discussed by Lee and Lee in reference 42.

107. Ahmad, N.; Srinivas, V. R.; Reddy, G. B.; Surolia, A., Thermodynamic Characterization of the Conformational Stability of the Homodimeric Protein, Pea Lectin. *Biochem.* **1998**, *37*, 16765–16772.

108. Dam, T. K.; Roy, R.; Das, S. K.; Oscarson, S.; Brewer, C. F., Binding of Multivalent Carbohydrates to Concanavalin A and *Dioclea Grandiflora* Lectin. *J. Biol. Chem.* **2000**, *275*, 14223-14230.

109. Dam, T. K.; Roy, R.; Page, D.; Brewer, C. F., Thermodynamic Binding Parameters of Individual Epitopes of Multivalent Carbohydrates to Concanavalin A As Determined by "Reverse" Isothermal Titration Microcalorimetry. *Biochem.* **2002**, *41*, 1359–1363.

110. Wiseman, T.; Williston, S.; Brandt, J. F.; Lin, L. N., Rapid Measurement of Binding Constants and Heats of Binding Using a New Titration Calorimeter. *Anal. Biochem.* **1989**, *179*, 131-135.

111. Khan, M. I.; Mandal, D. K.; Brewer, C. F., Interaction of Concanavalin A with Glycoproteins. A Quantitative Precipitation Study of Concanavalin A with the Soybean Agglutinin. *Carbohydr. Res.* **1991**, *213*, 69-77.

112. Tsutsumiuchi, K.; Aoi, K.; Okada, M., Ion Complex Formation between Poly(amido amine) Dendrimer HCl Salt and Poly(L-glutamic acid) Sodium Salt. *Polym. J.* **2000**, *32*, 107–112.

113. Tsuchida, E.; Abe, K., Interactions between Macromolecules in Solution and Intermacromolecular Complexes. *Adv. Polym. Sci.* **1982**, *45*, 1–119.

114. Osawa, T.; Matsumoto, I., Gorse (Ulex europeus) phytohemagglutinins. *Meth. Enzymol.* **1972**, *28*, 323–327.

115. Bhattacharyya, L.; Brewer, C. F., Preparation and Properties of Metal Ion Derivatives of the Lentil and Pea Lectins. *Biochem.* **1985**, *24*, 4974–4980.

116. Dam, T. K.; Brewer, C. F., Thermodynamic Studies of Lectin-Carbohydrate Interactions by Isothermal Titration Calorimetry. *Chem. Rev.* **2002**, *102*, 387-429.

117. Rich, R. L.; Myszka, D. G., Survey of the Year 2007 Commercial Optical Biosensor Literature. *J. Mol. Recognit.* **2008**, *21*, 355-400.

118. Rich, R. L. *et al.*, A Global Benchmark Study Using Affinity-Based Biosensors. *Anal. Biochem.* **2009**, *386*, 194-216.

119. Wilson, W. D., Analyzing Biomolecular Interactions. *Science* **2002**, *295*, 2103-2105.

120. Vornholt, W.; Hartmann, M.; Keusgen, M., SPR Studies of Carbohydrate-Lectin Interactions as Useful Tool for Screening on Lectin Sources. *Biosens. Bioelectron.* **2007**, *22*, 2983-2988.

121. Maljaars, C. E. P.; de Souza, A. C.; Halkes, K. M.; Upton, P. J.; Reeman, S. M.; Andre, S.; Gabius, H.-J.; McDonnell, M. B.; Kamerling, J. P., The Application of Neoglycopeptides in the Development of Sensitive Surface Plasmon Resonance-Based Biosensors. *Biosens. Bioelectron.* **2008**, *24*, 60-65.

122. Furuike, T.; Sadamoto, R.; Niikura, K.; Monde, K.; Sakairi, N.; Nishimura, S.-I., Chemical and Enzymatic Synthesis of Glycocluster Having Seven Sialyl Lewis X Arrays Using β-Cyclodextrin as a Key Scaffold Material. *Tetrahedron* **2005**, *61*, 1737-1742.

123. Suda, Y.; Arano, A.; Fukui, Y.; Koshida, S.; Wakao, M.; Nishimura, T.; Kusumoto, S.; Sobel, M., Immobilization and Clustering of Structurally Defined Oligosaccharides for Sugar Chips: An Improved Method for Surface Plasmon Resonance Analysis of Protein-Carbohydrate Interactions. *Bioconj. Chem.* **2006**, *17*, 1125-1135.

124. Tabarani, G.; Reina, J. J.; Ebel, C.; Vives, C.; Lortat-Jacob, H.; Rojo, J.; Fieschi, F., Mannose Hyperbranched Dendritic Polymers Interact with Clustered Organization of DC-SIGN and Inhibit gp120 Binding. *FEBS Lett.* **2006**, *580*, 2402-2408.

125. Mann, D. A.; Kanai, M.; Maly, D. J.; Kiessling, L. L., Probing Low Affinity and Multivalent Interactions with Surface Plasmon Resonance: Ligands for Concanavalin A. *J. Am. Chem. Soc.* **1998**, *120*, 10575-10582.

126. Munoz, E. M.; Correa, J.; Fernandez-Megia, E.; Riguera, R., Probing the Relevance of Lectin Clustering for the Reliable Evaluation of Multivalent Carbohydrate Recognition. *J. Am. Chem. Soc.* **2009**, *131*, 17765-17767.

127. Kensinger, R. D.; Yowler, B. C.; Benesi, A. J.; Schengrund, C.-L., Synthesis of Novel, Multivalent Glycodendrimers as Ligands for HIV-1 gp120. *Bioconj. Chem.* **2004**, *15*, 349-358.

128. Cecioni, S.; Lalor, R.; Blanchard, B.; Praly, J.-P.; Imberty, A.; Matthews, S. E.; Vidal, S., Achieving High Affinity Towards a Bacterial Lectin Through Multivalent Topological Isomers of Calix[4]arene Glycoconjugates. *Chem. Eur. J.* **2009**, *15*, 13232-13240.

129. Nuzzo, R. G.; Allara, D. L., Adsorption of Bifunctional Organic Disulfides on Glod Surfaces. *J. Am. Chem. Soc.* **1983**, *105*, 4481-4483.

130. Bain, C. D.; Troughton, E. B.; Tao, Y.-T.; Evall, J.; Whitesides, G. M.; Nuzzo, R. G., Formation of Monolayer Films by the Spontaneous Assembly of Organic Thiols from Solution onto Gold. *J. Am. Chem. Soc.* **1989**, *111*, 321-335.

131. Chechik, V.; Crooks, R. M.; Stirling, C., Reactions and Reactivity in Self-Assembled Monolayers. *Adv. Mater.* **2000**, *12*, 1161-1171.

132. Fryxell, G. E.; Rieke, P. C.; Wood, L. L.; Engelhard, M. H.; Williford, R. E.; Graff, G. L.; Campbell, A. A.; Wiacek, R. J.; Lee, L.; Halverson, A., Nucleophilic Displacements in Mixed Self-Assembled Monolayers. *Langmuir* **1996**, *12*, 5064-5075.

133. Weck, M.; Jackiw, J. J.; Rossi, R. R.; Weiss, P. S.; Grubbs, R. H., Ring-Opening Metathesis Polymerization from Surfaces. *J. Am. Chem. Soc.* **1999**, *121*, 4088-4089.

134. Bartz, M.; Küther, J.; Seshadri, R.; Tremel, W., Colloid-Bound Catalysts for Ring-Opening Metathesis Polymerization: A Combination of Homogenous and Heterogenous Properties. *Angew. Chem. Int. Ed.* **1998**, *37*, 2466-2468.

135. Bain, C. D.; Biebuyck, H. A.; Whitesides, G. M., Comparison of Self-Assembled Monolayers on Gold: Coadsorption of Thiols and Disulfides. *Langmuir* **1989**, *5*, 723-727.

136. Lahiri, J.; Isaacs, L.; Tien, J.; Whitesides, G. M., A Strategy for the Generation of Surfaces Presenting Ligands for Studies of Binding Based on an Active Ester as a Common Reactive Intermediate: A Suface Plasmon Resonance Study. *Anal. Chem.* **1999**, *71*, 777-790.

137. Schlenoff, J. B.; Li, M.; Ly, H., Stability and Self-Exchange in Alkanethiol Monolayers. *J. Am. Chem. Soc.* **1995**, *117*, 12528-12536.

138. Subramanian, A.; Irudayaraj, J.; Ryan, T., Mono and Dithiol Surfaces on Surface Plasmon Resonance Biosensors for the Detection of *Staphylococcus Aureus*. *Sensors and Actuators B* **2006**, *114*, 192-198.

139. Zhao, Y.; Perez-Segarra, W.; Shi, Q.; Wei, A., Dithiocarbamate Assembly on Gold. *J. Am. Chem. Soc.* **2005**, *127*, 7328-7329.

140. Mandal, D. K.; Kishore, N.; Brewer, C. F., Thermodynamics of Lectin-Carbohydrate Interactions. Titration Microcalorimetry Measurements of the Binding of N-Linked Carbohydrates and Ovalbumin to Concanavalin A. *Biochemistry* **1994**, *33*, 1149-1156.

141. Walter, E. D.; Sebby, K. B.; Usselman, R. J.; Singel, D. J.; Cloninger, M. J., Characterization of Heterogenously Functionalized Dendrimers by Mass Spectrometry and EPR Spectroscopy. *J. Phys. Chem. B* **2005**, *109*, 21532-21538.

142. Chervenak, M. C.; Toone, E. J., Calorimetric Analysis of the Binding of Lectins with Overlapping Carbohydrate-Binding Ligand Specificities. *Biochemistry* **1995**, *34*, 5685-5695.

143. Dimick, S. M.; Powell, S. C.; McMahon, S. A.;Moothoo, D. N., Naismith, J. H.; Toone, E. J., On the Meaning of Affinity: Cluster Glycoside Effects and Concanavalin A. *J. Am. Chem. Soc.* **1999**, *121*, 10286-10296.

144. De la Fuente, J. M.; Eaton, P.; Barrientos, A. G.; Menedez, M.; Penades, S., Thermodynamic Evidence for Ca^{2+}-Mediated Self-Aggregation of Lewis X Gold Glyconanoparticles. A Model for Cell Adhesion via Carbohydrate-Carbohydrate Interaction. *J. Am. Chem. Soc.* **2005**, *127*, 6192-6197.

145 Santacroce, P. V.; Basu, A., Studies of the Carbohydrate-Carbohydrate Interaction Between Lactose and GM_3 Using Langmuir Monolayers and Glycolipid Micelles. *Glycoconj. J.* **2004**, *21*, 89-95.

146. Motulsky, H. J.; Christopoulos, A., *Fitting Models to Biological Data Using Linear and Nonlinear Regression. A Practical Guide to Curve Fitting.* 2003, GraphPad Software Inc., San Diego CA, www.graphpad.com.

147. Lindhorst, T. K.; Kotter, S.; Krallmann-Wenzel, U.; Ehlers, S., Trivalent α-D-Mannoside Clusters as Inhibitors of Type-1 Fimbriae-Mediated Adhesion of *Escherichia Coli*: Structural Variation and Biotinylation. *J. Chem. Soc. Perkin Trans 1* **2001**, 823-831.

148. Fukuda, M.; Hingsgaul, O., *Molecular and Cellular Glycobiology* Oxford University Press: Oxford, 2000.

149. Chen, S.; Zheng, T.; Shortreed, M. R.; Alexander, C.; Smith, L. M., Analysis of Cell Surface Carbohydrate Expression Patterns in Tumorigenic Human Breast Cell Lines Using Lectin Arrays. *Anal. Chem.* **2007**, *79*, 5698-5702.

150. Bertozzi, C. R.; Kiessling, L. L., Chemical Glycobiology. *Science* **2001**, *291*, 2357-2364.

151. Lin, S.; Kemmner, W.; Grigull, S.; Schlag, P. M., Cell Surface α2,6-Sialylation Affects Adhesion of Breast Carcinoma Cells. *Exp. Cell Res.* **2002**, *276*, 101-110.

152. Hakomori, S., Tumor Malignancy Defined by Aberrant Glycosylation and Sphingo(glyco)lipid Metabolism. *Cancer Res.* **1996**, *56*, 5309-5318.

153. Dansey, R.; Murray, J.; Ninin, D.; Bezwoda, W. R., Lectin Binding in Human Breast Cancer: Clinical and Pathologic Correlations with Fluorescein-conjugated Peanut, Wheat Germ and Concanavalin A Binding. *Oncology* **1988**, *45*, 300-302.

154. Al-Mughaid, H.; Grindley, T. B., Synthesis of a Nonavalent Mannoside Glycodendrimer Based on Pentaerythriol. *J. Org. Chem.* **2006**, *71*, 1390-1398.

155. Blanchfield, J.; Toth, I., Lipid, Sugar and Liposaccharide Based Delivery Systems 2. *Curr. Med. Chem.* **2004**, *11*, 2375-2382.

156. Patel, A.; Lindhorst, T. K., Multivalent Glycomimetics: Synthesis of Nonavalent Mannoside Clusters with Variation of Spacer Properties. *Carbohyd. Res.* **2006**, *341*, 1657-1668.

157. Köhn, M.; Benito, J. M.; Mellet, C. O.; Lindhorst, T. K.; Fernández, J. M. G., Functional Evaluation of Carbohydrate-Centred Glycoclusters by Enzyme-Linked Lectin Assay: Ligands for Concanavalin A. *ChemBioChem* **2004**, *5*, 771.

158. Dubber, M.; Lindhorst, T. K., Exploration of Reductive Amination for the Synthesis of Cluster Glycosides. *Synthesis-Stuttgart* **2001**, 327-330.

159. Kötter, S.; Krallmann-Wenzel, U.; Ehlers, S.; Lindhorst, T. K., Multivalent Ligands for the Mannose-Specific Lectin on Type 1 Fimbriae of *Escherichia Coli*: Syntheses and Testing of Trivalent α-D-Mannoside Clusters. *J. Chem. Soc., Perkin Trans. 1* **1998**, 2193-2200.

160. Kikkeri, R.; Garcia-Rubio, I.; Seeberger, P. H., Ru(II)-Carbohydrate Dendrimers as Photoinduced Electron Transfer Lectin Biosensors. *Chem. Commun.* **2009**, 235-237.

161. Hasegawa, T.; Yonemura, T.; Mastuura, K.; Kobayashi, K., Tris-Bipyridine Ruthenium Complex-Based Glycoclusters: Amplified Luminescence and Enhanced Lectin Affinities. *Tet. Lett.* **2001**, *42*, 3989-3992.

162. Sato, K.; Hada, N.; Takeda, T., Syntheses of New Peptidic Glycoclusters Derived from β-Alanine: Di- and Trimerized Glycocluster-Clusters. *Carbohyd. Res.* **2006**, *341*, 836-845.

163. Ozaki, K.; Lee, R. T.; Lee, Y. C.; Kawasaki, T., The Differences in Structural Specificity for Recognition and Binding Between Asialoglycoprotein Receptors of Liver and Macrophages. *Glycoconj. J.* **1995**, *122*, 268-274.

164. Kolb, H. C.; Finn, M. G.; Sharpless, K. B., Click Chemistry: Diverse Chemical Function from a Few Good Reactions. *Angew. Chem. Int. Ed.* **2001**, *40*, 2004-2021.

165. Kolb, H. C.; Sharpless, K. B., The Growing Impact of Click Chemistry on Drug Discovery. *Drug Discovery Today* **2003**, *8*, 1128-1137.

166. Gouin, S. G.; Wellens, A.; Bouckaert, J.; Kovensky, J., Synthetic Multimeric Heptyl Mannosides as Potent Antiadhesives of Uropathogenic Escherichia Coli. *ChemMedChem* **2009**, *4*, 749-755.

167. Chen, Y.; Zhao, L.; Huang, Z.; Zhao, Y.; Li, Y., Facile Synthesis of Cyclopeptide-Centered Multivalent Glycoclusters with 'Click Chemistry' and Molecular Recognition Study by Surface Plasmon Resonance. *Bioorg. Med. Chem. Lett.* **2009**, *19*, 3775-3778.

168. Martin, A.; Li, B.; Gillies, E. R., Surface Functionalization of Nanomaterials with Dendritic Groups: Toward Enhanced Binding to Biological Targets. *J. Am. Chem. Soc.* **2008**, *131*, 734-741.

169. Chabre, Y. M.; Contino-Pépin, C.; Placide, V.; Shiao, T. C.; Roy, R., Expeditive Synthesis of Glycodendrimer Scaffolds Based on Versatile TRIS and Mannoside Derivatives. *J. Org. Chem.* **2008**, *73*, 5602-5605.

170. Touaibia, M.; Shiao, T. C.; Papadopoulos, A.; Vaucher, J.; Wang, Q.; Benhamioud, K.; Roy, R.. Tri- and Hexavalent Mannoside Clusters as Potential Inhibitors of Type 1 Fimbriated Bacteria Using Pentaerythritol and triazole Linkages. *Chem. Commun.* **2007**, 380-382.

171. Touaibia, M.; Wellens, A.; Shiao, T. C.; Wang, Q.; Sirois, S.; Bouckaert, J.; Roy, R., Mannosylated G(0) Dendrimers with Nanomolar Affinities to Escherichia Coli FimH. *ChemMedChem* **2007**, *2*, 1190-1201.

172. Deyl, Z., *Electrophoresis: A Survey of Techniques and Applications*. Elsevier Scientific: Amsterdam/New York, 1983.

173. Jay, G. D.; Culp, D. J.; Jahnke, M. R., Silver Staining of Extensively Glycosylated Proteins on Sodium Dodecyl Sulfate-Polyacrylamide Gels: Enhancement by Carbohydrate-binding Dyes. *Anal. Biochem.* **1990**, *185*, 324-330.

174. Zacharius, R. M.; Zell, T. E.; Morrison, J. H.; Woodlock, J. J., Glycoprotein Staining Following Electrophoresis on Acrylamide Gels. *Anal. Biochem.* **1969**, *30*, 148-152.

175. Lis, H.; Sharon, N., Lectins: Carbohydrate-Specific Proteins That Mediate Cellular Recognition. *Chem. Rev.* **1998**, *98*, 637-674.

176. Dzantiev, L.; Alekseyev, Y. O.; Morales, J. C.; Kool, E. T.; Romano, L. J., Significance of Nucleobase Shape Complementarity and Hydrogen Bonding in the Formation and Stability of the Closed Polymerase-DNA Complex. *Biochemistry* **2001**, *40*, 3215-3221.

177. Norel, R.; Petrey, D.; Wolfson, H. J.; Nussinov, R., Examination of Shape Complementarity in Docking of Unbound Proteins. *Proteins* **1999**, *36*, 307-317.

178. Bernstein, F. C.; Koetzle, T. F.; Williams, G. J. B.; Meyer Jr., E. F.; Brice, M. D.; Rodgers, J. R.; Kennard, O.; Shimanouchi, T.; Tasumi, M., The Protein Data Bank: A Computer-Based Archival File for Macromolecular Structures. *J. Mol. Biol.* **1977**, *112*, 535-542.

179. Goldstein, I. J.; Poretz, R. D., In *The Lectins*; Liener, I. E., Sharon, N., Goldstein, I. J., Eds.; Academic Press: New York, 1986; p 35

180. Weatherman, R. V.; Kiessling, L. L., Fluorescence Anisotropy Assays Reveal Affinities of C- and O-Glycosides for Concanavalin A. *J. Org. Chem.* **1996**, *61*, 534-538.

181. Nangia-Makker, P.; Balan, V.; Raz, A., Regulation of Tumor Progression by Extracellular Galectin-3. *Cancer Microenviron.* **2008**, *1*, 43-51.

182. Wood, S. J.; Maleeff, B.; Hart, T.; Wetzel, R., Physical, Morphological and Functional Differences Between pH 5.8 and 7.4 Aggregates of Alzheimer's Amyloid Peptide Aβ. *J. Mol. Biol.* **1996**, *256*, 870-877.

183. Segal, D. M.; Weiner, G. J.; Weiner, L. M., Bispecific Antibodies in Cancer Therapy. *Curr. Opin. Immunol.* **1999**, *11*, 558-562.

184. Singh, R. S.; Tiwary, A. K.; Kennedy, J. F., Lectin: Sources, Activities, and Applications. *Crit. Rev. Biotechnol.* **1999**, *19*, 145-178.

185. Gestwicki, J. E.; Strong, L. E.; Cairo, C. W.; Boehm, F. J.; Kiessling, L. L., Cell Aggregation by Scaffolded Receptor Clusters. *Chem. Biol.* **2002**, *9*, 163-169.

186. Satrijo, A.; Swager, T. M., Anthryl-Doped Conjugated Polyelectrolytes as Aggregation-Based Sensors for Nonquenching Multicationic Analytes. *J. Am. Chem. Soc.* **2007**, *129*, 16020-16028.

187. Maynor, M. S.; Nelson, T. L.; O'Sullivan, C.; Lavigne, J. J., A Food Freshness Sensor Using the Multistate Response from Analyte-Induced Aggregation of a Cross-Reactive Poly(thiopene). *Org. Lett.* **2007**, *9*, 3217-3220.

188. Liu, B.; Pu, K., A Multicolor Cationic Conjugated Polymer for Naked-Eye Detection and Quantification of Heparin. *Macromolecules* **2008**, *41*, 6636-6640.

189. Wang, M.; Zhang, D.; Zhang, G.; Tang, Y.; Wang, S.; Zhu, D., Fluorescence Turn-On Detection of DNA and Label-Free Fluorescence Nuclease Assay Based on the Aggregation-Induced Emission of Silole. *Anal. Chem.* **2008**, *80*, 6443-6448.

190. Chen, K.; Yang, S.; Hwang, C.; Fang, J., Phospholipid-Induced Aggregation and Anthracene Excimer Formation. *Org. Lett.* **2008**, *10*, 4401-4404.

191. Lakowicz, J. R., *Principles of Fluorescence Spectroscopy*. Kluwer Academic: New York, 1999.

192. Rao, C.V. *Immunology: A Textbook* First Edition, Alpha Science Intl Ltd: Oxford, UK, 2005.

193. Dam, T. K.; Gerken, T. A.; Brewer, C. F., Thermodynamics of Multivalent Carbohydrate-Lectin Cross-Linking Interactions: Importance of Entropy in the Bind and Jump Mechanism. *Biochemistry* **2009**, *48*, 3822-3827.

194. Horan, N.; Yan, L.; Isobe, H.; Whitesides, G. M.; Kahne, D., Nonstatistical Binding of a Protein to Clustered Carbohydrates. *Proc. Natl. Acad. Sci.* **1999**, *96*, 11782-11786.

195. Halligan, B. D.; Ruotti, V.; Jin, W.; Laffoon, S.; Twigger, S. N.; Dratz, E. A., ProMoST (Protein Modification Screening Tool): A Web-Based Tool for Mapping Protein Modifications on Two-Dimensional Gels. *Nucleic Acids Res.* **2004**, *32*, W638-W644.

196. Davis, A. P.; Wareham, R. S., Carbohydrate Recognition through Noncovalent Interactions: A Challenge for Biomimetic and Supramolecular Chemistry. *Angw. Chem. Int. Ed.* **1999**, *38*, 2978-2996.

197. Wang, W.; Gao, X.; Wang, B., Boronic Acid-Based Sensors. *Curr. Org. Chem.* **2002**, *6*, 1285-1317.

198. James, T. D.; Shinkai, S., Artificial Receptors as Chemosensors for Carbohydrates. *Top. Curr. Chem.* **2002**, *218*, 159-200.

199. Yang, W.; Fan, H.; Gao, X.; Gao, S.; Karnati, V. V. R.; Ni, W.; Hooks, W. B.; Carson, J.; Weston, B.; Wang, B., The First Fluorescent Diboronic Acid Sensor Specific

for Hepatocellular Carcinoma Cells Expressing Sialyl Lewis X. *Chem. Biol.* **2004**, *11*, 439-448.

200. Dowlut, M.; Hall, D. G., An Improved Class of Sugar-Binding Boronic Acids, Soluble and Capable of Complexing Glycosides in Neutral Water. *J. Am. Chem. Soc.* **2006**, *128*, 4226-4227.

201. Galardy, R. E.; Craig, L. C.; Jameison, J. D.; Printz, M. P., Photoaffinity Labelling of Peptide Hormone Binding Sites. *J. Biol. Chem.* **1974**, *249*, 3510-3518.

202. Dormán, G.; Prestwich, G. D., Benzophenone Photophores in Biochemistry. *Biochemistry* **1994**, *33*, 5661-5673.

203. Spicka, K. J., *Design and Synthesis of Fluorescent Dyes for Use in Proteomic Research.* **2008**, Ph. D. Thesis, Montana State University.

204. Unpublished results from the laboratory of professor Paul A. Grieco

205. Springsteen, G.; Wang, B., A Detailed Examination of Boronic Acid-Diol Complexation. *Tetrahedron* **2002**, 5291-5300.

206. Zhang, Y.; Zhang, Z., Low-Affinity Binding determined by Titration Calorimetry Using a High-Affinity Coupling Ligand: a Thermodynamic Study of Ligand Binding to Protein Tyrosine Phosphatase 1B. *Anal. Biochem.* **1998**, *261*, 139-148.

207. Spiro, R. G., Studies on Fetuin, a Glycoprotein of Fetal Serum. I. Isolation, Chemical Composition and Physiochemical Properties. *J. Biol. Chem.* **1960**, *235*, 2860-2869.

208. Unpublished results from the laboratory of professor Paul A. Grieco

CPSIA information can be obtained
at www.ICGtesting.com
Printed in the USA
LVIW020802290513
335930LV00005B